Palgrave Studies in Climate Resilient Societies

Series Editor
Robert C. Brears
Avonhead, Canterbury, New Zealand

The Palgrave Studies in Climate Resilient Societies series provides readers with an understanding of what the terms **resilience and climate resilient** societies mean; the best practices and lessons learnt from various governments, in both non-OECD and OECD countries, implementing climate resilience policies (in other words what is 'desirable' or 'undesirable' when building climate resilient societies); an understanding of what a resilient society potentially looks like; knowledge of when resilience building requires slow transitions or rapid transformations; and knowledge on how governments can create coherent, forward-looking and flexible policy innovations to build climate resilient societies that: support the conservation of ecosystems; promote the sustainable use of natural resources; encourage sustainable practices and management systems; develop resilient and inclusive communities; ensure economic growth; and protect health and livelihoods from climatic extremes.

Smith I Azubuike
Ayodele Asekomeh • Obindah Gershon
Editors

Decarbonisation Pathways for African Cities

palgrave
macmillan

Editors
Smith I Azubuike
Durham Law School
Durham University
Durham, UK

Obindah Gershon
Centre for Economic Policy and
Development Research (CEPDeR)
Department of Economics and
Development Studies
Covenant University
Ota, Nigeria

Eduardo Mondlane University
Maputo, Mozambique

Ayodele Asekomeh
Department of Accounting
and Finance
Aberdeen Business School
Robert Gordon University
Aberdeen, UK

ISSN 2523-8124 ISSN 2523-8132 (electronic)
Palgrave Studies in Climate Resilient Societies
ISBN 978-3-031-14005-1 ISBN 978-3-031-14006-8 (eBook)
https://doi.org/10.1007/978-3-031-14006-8

This Palgrave Macmillan imprint is published by the registered company Springer Nature Switzerland AG.
The registered company address is: Gewerbestrasse 11, 6330 Cham, Switzerland

Contents

1 **Introduction: Decarbonising African Cities in a Carbon-Constrained World** 1
Smith I. Azubuike, Obindah Gershon, and Ayodele Asekomeh

Part I **Decarbonising African Cities: Strategies and Applications** 13

2 **Solar Urban Planning in African Cities: Challenges and Prospects** 15
Mark M. Akrofi, Mahesti Okitasari, Olayinka S. Ohunakin, and Smith I. Azubuike

3 **Contextualising Waste Management Operations Towards Low-Carbon African Cities** 37
Laura Muniafu and Nzembi Mutiso

4 **Innovative Strategies for Decarbonising the Healthcare Sector in Nigerian Cities** 51
Smith I. Azubuike and Adebola Adeyemi

5 Optimising Hybrid Power Systems for Sustainable Operation of Remote Telecommunication Infrastructure 73
Michael S. Okundamiya and Samuel T. Wara

6 Performance Analysis of a Grid-Linked Microgrid System in a University Campus 95
Samuel T. Wara and Michael S. Okundamiya

Part II Governance and Policy Approaches for Decarbonising African Cities 115

7 Powering Action Towards Energising African Cities Sustainably: Perspectives from Kenya 117
Laura Muniafu, Peter Ombewa, and Nzembi Mutiso

8 The Political Economy of Decarbonising African Petro-cities: Governance Reconfigurations for the Future 135
Magnus C. Abraham-Dukuma, Okechukwu C. Aholu, Jesse Nyokabi, and Michael O. Dioha

9 To Opt-in or to Cop Out: COP26 and the Policy Dynamics of Decarbonising African Cities 157
Ayodele Asekomeh, Obindah Gershon, and Smith I. Azubuike

10 Conclusion: Towards a Decarbonisation Framework for African Cities 179
Obindah Gershon, Smith I. Azubuike, and Ayodele Asekomeh

Index 187

Notes on Contributors

Magnus C. Abraham-Dukuma is a Senior Policy Advisor on Climate Change at the Science and Strategy Hub, Gisborne District Council, New Zealand. His research interests traverse energy law and policy, climate law and policy, energy justice, and just transition to a climate-resilient future.

Adebola Adeyemi is Assistant Professor in Law at Durham Law, Durham University. Adeyemi teaches private law and researches energy law, sustainable project finance, and IT law. Demonstrating his interest in sustainability, Adeyemi sits on the Board of FREEE Recycle Holding Limited UK.

Okechukwu C. Aholu is an Associate Lecturer at the University of West of England. He is also the Deputy Director of the African Natural Resources and Energy Law Network and a researcher with Escrow Tech Limited. His research interests are Energy Law, Environmental Law, and International Trade law.

Mark M. Akrofi is a PhD candidate in Sustainability Science at the United Nations University, Institute for the Advanced Study of Sustainability in Japan. He holds a Master's degree in Energy Policy and specializes in researching governance and policy aspects of sustainable energy transitions in Africa.

Ayodele Asekomeh is Senior Lecturer at the Aberdeen Business School, Robert Gordon University. A multiple Chartered Accountant with Big-4 experience, his work considers the interactions and discourses among state actors, investors and other stakeholders on financing, accountability, governance, sustainability, and risk in the energy and extractive industries in the context of climate action.

Smith I. Azubuike is Assistant Professor in Energy Law at Durham Law School, Durham University and a qualified Lawyer in Nigeria. He previously taught at Queen's University Belfast. Smith's teaching and research focus on energy law and sustainability, risk allocation in the energy sector and renewable energy law. He is the Book Reviews Editor of Edinburgh University Press' Global Energy Law and Sustainability journal.

Michael O. Dioha is a Postdoctoral Research Fellow in the Carnegie Institution for Science's Department of Global Ecology at Stanford. His research interests revolve around energy-environment-economic modelling, the quantification of energy/emissions scenarios, and interdisciplinary issues in energy and climate justice. Dioha is a Fellow of the Energy for Growth Hub.

Obindah Gershon holds a PhD in economics. He Chairs the Centre for Economic Policy and Development Research (CEPDeR) and is a Senior Lecturer at the Department of Economics and Development Studies Covenant University, Nigeria. He is a Visiting Professor at Eduardo Mondlane University, Mozambique and a member of the IAEE.

Laura Muniafu is the Director at Strathmore Extractives Industry Centre (SEIC), where she is involved in research, capacity building, policy development and stakeholder engagement in Africa's extractives industry. She holds a B.Eng in Chemical Engineering from Sheffield University and MSc. in Oil and Gas Engineering from Aberdeen University.

Jesse Nyokabi works as Control Engineer in a power plant. Through Green Energy Pacesetter, he is also a champion for green energy. He explores innovative opportunities to accelerate the impact of green energy on the environment, economy, and communities. He is pursuing postgraduate research in supercritical geothermal drilling.

Nzembi Mutiso is a Researcher at the Extractives Baraza, Nairobi. She is a lawyer with experience in research, advocacy, and project management. Nzembi has participated in various community mobilization, awareness, and advocacy initiatives, including legal aid clinics, international solidarity campaigns on human rights, and lobbying for legislation and policies.

Olayinka S. Ohunakin is a Senior Lecturer at the Energy and Environment Research Group (TEERG), Mechanical Engineering Department, Covenant University, Ogun State, Nigeria. He is also a Senior Research Associate, Faculty of Engineering & the Built Environment, University of Johannesburg, South Africa.

Mahesti Okitasari is a PhD candidate at the Institute for the Advanced Study of Sustainability, United Nations University, 5-53-70 Jingumae, Shibuya-ku, Tokyo, 150-8925, Japan.

Michael S. Okundamiya is an Associate Professor at Ambrose Alli University, Nigeria. He has over 18 years of international experience in independent and collaborative research in diverse areas of Electrical, Electronics & Computer Engineering. His current interests include power electronics, energy systems integration, communication technologies, and soft computing.

Peter Ombewa is a Hydraulics Application Engineer at Danfoss (formerly Eaton Corp) Kenya. He has a wealth of experience in the energy and petroleum sectors. He holds a bachelor's degree in Mechanical Engineering from the University of Nairobi and a master's degree in Oil & Gas Engineering from the University of Aberdeen.

Samuel T. Wara is Vice-Chancellor at Havilla University, Nde-Ikom, Nigeria. He is a Fellow of the Nigerian Society of Engineers and a Chartered Electrical Engineer, a teacher, researcher, manager, and administrator. He researched and lectured on Renewable Energy, Environment, Energy Conservation & Efficiency, Power quality and Distributed Energy.

List of Figures

Fig. 2.1	Electricity access in urban areas as a percentage of the urban population (blue), and the urban population as a percentage of the total population (red) for sub-Saharan African countries with less than 80% electricity access in urban areas (2014). Source: IRENA (2019)	17
Fig. 2.2	Solar urban planning process. Source: Amado and Poggi (2012)	19
Fig. 2.3	Document identification and selection process. Adapted from Page et al. (2021)	21
Fig. 2.4	Characteristics of selected documents: types of documents (**a**), trend of publications (**b**), regional focus of publications (**c**), and the number of publications by country (**d**)	23
Fig. 2.5	Annual solar PV energy yield for small residential (**a**) and medium commercial (**b**) PV systems in selected African cities. Source: Authors' construct based on data from the World Bank (2020)	24
Fig. 2.6	Global weighted average total installed costs, capacity factors, and LCOE for solar PV, 2010–2019. Source: IRENA (2020)	25
Fig. 2.7	SWOT analysis	30
Fig. 2.8	Aerial image of a gated community in Accra, Ghana (picture taken by the authors)	31
Fig. 4.1	The impact of climate change on human health (Source: Health Care Without Harm 2019)	60
Fig. 5.1	Architecture of HPS for off-grid cellular sites	78
Fig. 5.2	Load profiles of the 4G LTE eNode B 2/2/2 infrastructure under study	84
Fig. 5.3	Effect of each subsystem on the total cost of WT/PV/Battery Hybrid Power System	85

Fig. 5.4	Annual electric power output of subsystem generators (**a**) WT (**b**) PV	86
Fig. 5.5	SOC of the battery system	86
Fig. 5.6	Comparison of cumulative cash flow of the baseline and proposed systems over the project lifespan	88
Fig. 5.7	Influence of inflation on the economy of the optimised WT/PV/Battery system at a discount rate of 8%	89
Fig. 6.1	Design of grid-linked PV/FC/diesel hybrid system for ICT load	99
Fig. 6.2	Simulated grid electricity distribution for study site	100
Fig. 6.3	Fundamental constituents of water electrolysers fragmented at different levels (IRENA 2020)	101
Fig. 6.4	Fuel and efficiency curves of DG used in this study	104
Fig. 6.5	Analysis of the typical daily electric load profile of the ICT infrastructure under study	105
Fig. 6.6	Electric load profiles of ICT infrastructure under study (**a**) seasonal (**b**) yearly	106
Fig. 6.7	Cumulative cash flow of the baseline and proposed systems compared over project lifespan	109
Fig. 8.1	Future scenarios hierarchical pyramid for African petro-cities. Source: Authors	142
Fig. 8.2	Governance reconfigurations for decarbonising African petro-cities. Source: Authors	145
Fig. 9.1	Annual CO2 emissions from fossil fuels, by world region [Carbon dioxide (CO2) emissions from the burning of fossil fuels for energy and cement production. Land use change is not included] (Source: Global Carbon Project—Our World in Data [https://ourworldindata.org/co2-and-other-greenhouse-gas-emissions] CC BY)	161
Fig. 9.2	Per capita CO2 emissions, 2020 [Carbon dioxide (CO2) emissions from the burning of fossil fuels for energy and cement production. Land use change is not included] (Source: Global Carbon Project—Our World in Data [https://ourworldindata.org/co2-and-other-greenhouse-gas-emissions] CC BY)	162
Fig. 9.3	The Advocacy Coalition Framework (Source: Sabatier and Weible 2007, p.202)	165

List of Tables

Table 4.1	Energy use from Cedarcrest Hospitals, Abuja, Nigeria	59
Table 5.1	Evolution of site topology of cellular generation	76
Table 5.2	The techno-economic description of the chosen WT (IRENA 2021; https://www.windpowercn.com/products/31.html)	79
Table 5.3	Techno-economic description of the selected battery (IRENA 2021; https://www.bae-berlin.de/)	80
Table 5.4	Description of converter/generator used in the hybrid system design (Okundamiya 2021a; Ghenai and Bettayeb 2019)	81
Table 5.5	Monthly averages for meteorological data for Maiduguri (Okundamiya et al. 2014b; NASA (National Aeronautics and Space Administration), Prediction of Worldwide Energy Resource (POWER) database)	82
Table 5.6	Characteristics of meteorological datasets collected from NIMET (Okundamiya et al. 2014b)	83
Table 5.7	Monetary cost details of the subsystem constituents for proposed WT/PV/battery hybrid power system for4G LTE eNode B 2/2/2 Cellular Site	85
Table 5.8	Comparison of techno-economy of WT/PV/Battery (Proposed) and DG/Battery/Converter (Baseline) Hybrid Systems	87
Table 6.1	Constraints for the erratic grid electricity distribution design	100
Table 6.2	Cost details of modules used in the FC system design (Okundamiya 2021b; IEA 2019)	103
Table 6.3	Economic description of DG utilised in the design (Okundamiya et al. 2022)	104

Table 6.4	22-Year monthly averages for global horizontal radiation for study site (NASA (National Aeronautics and Space Administration), Prediction of Worldwide Energy Resource (POWER) database)	105
Table 6.5	Economic description of power converter/controller used in system design (Okundamiya et al. 2022)	106
Table 6.6	Comparison of architecture of proposed design (grid/PV/FC/DG) and baseline (grid/DG) hybrid system	107
Table 6.7	Contribution of the individual subsystems to the overall annual electricity generation of the proposed hybrid system design (grid/PV/FC/DG)	107
Table 6.8	Comparison of techno-economic merits of proposed and baseline systems	108
Table 6.9	Comparison analysis of economics of proposed and baseline systems	109
Table 6.10	Pollutant emission characteristics of proposed and baseline systems	110
Table 8.1	Economic impact of oil on African OPEC member countries	139
Table 8.2	Dynamics of African petro-cities	140

CHAPTER 1

Introduction: Decarbonising African Cities in a Carbon-Constrained World

Smith I. Azubuike, Obindah Gershon, and Ayodele Asekomeh

Abstract This chapter presents arguments for addressing the key questions in the decarbonisation discourse. For instance, what determines the decarbonisation process, and how could low-carbon processes be implemented sustainably in African cities? Additionally, what are the governance and policy reconfigurations needed to support the investments towards reducing carbon emissions in African cities? So, towards answering these

S. I. Azubuike (✉)
Durham Law School, Durham University, Durham, UK

O. Gershon
Centre for Economic Policy and Development Research (CEPDeR), Department of Economics and Development Studies, Covenant University, Ota, Nigeria

Eduardo Mondlane University, Maputo, Mozambique

A. Asekomeh
Department of Accounting and Finance, Aberdeen Business School, Robert Gordon University, Aberdeen, UK

S. I. Azubuike et al. (eds.), *Decarbonisation Pathways for African Cities*, Palgrave Studies in Climate Resilient Societies,
https://doi.org/10.1007/978-3-031-14006-8_1

questions, two broad categorisation are proposed—(1) Decarbonising African Cities: Strategies and Applications; (2) Governance and Policy Approaches for Decarbonising African Cities. It further proposes that decarbonisation be approached from the specific context of sectors and integrated within an appropriate policy and governance framework for each city. Furthermore, institutions to effectively implement the policies are needed.

Keywords Decarbonisation • Governance and policy approaches • Africa • Policy change

1.1 Background

Fossil fuels and other carbon-emitting sources (firewood) and practices (bush burning etc.) in Africa are critical contributors to climate change (Nalule and Azubuike 2020). They also primarily offer economic rent in the region despite the environmental concerns about their use. Fankhauser and Jotzo (2018) note that fossil fuels are critical to the economic growth and development projections of resource-rich countries, especially in Africa. But the challenge brought about by climate change, and the discussions on global emissions reduction are mainstreaming the narrative for a decarbonised society. As such, several countries have pledged to fast-track and intensify actions needed for a sustainable low-carbon future through the ratification of the Paris Agreement. Achieving the Paris Agreement's objective of limiting global warming to 1.5 °C to combat climate change requires urgent and comprehensive action from states, regions, and at the international level (NEA 2019; Fay et al. 2015). This urgent action is necessary as global population growth increases both energy demand and the widespread use of carbon-emitting sources in millions of homes to meet this demand, especially in developing countries (Daramola et al. 2021).

Decarbonisation is a plausible platform for achieving the climate action plan, both in the energy and non-energy sectors. So far, the focus has been mainly on low-carbon transition in the energy sector, leaving out other critical areas and aspects of daily life that contribute to greenhouse gas (GHG) emissions in Africa. Climate-resilient buildings, smart metres, solar PV panels, eco-friendly transport systems, and low-carbon practices in waste management and other off-grid energy systems can attract local and foreign investments. There is currently limited investment in this area (Echeverri 2018). Amidst the ongoing energy transition, increasing urbanisation and rural-urban migration in Africa signify the need for creating decarbonised cities. Different strategies have been proposed, including limiting carbon-emitting sources and eliminating practices that contribute to greenhouse gas (GHG) emissions within cities Altieri et al., 2015. In this regard, Africa represents an opportunity for investment in low-carbon sources and has a market that can attract investors due to its population.

Africa is a densely populated continent with visible energy poverty and energy access challenges (Ugwoke et al. 2020; Hafner et al. 2018) despite the abundance of energy resources. Most houses are not energy-efficient or not designed to be climate-resilient. Insufficient or limited access to power from the national grid forces millions of Africans to utilise carbon-emitting or fossil fuel sources for cooking and lighting their homes which contributes to global warming (Gershon and Asaolu 2020; IPCC 2018). The International Energy Agency (IEA) notes that about two-thirds of Africa's population, equivalent to nearly 620 million people, do not have access to electricity and almost 730 million depend on traditional solid biomass for cooking (IEA 2019). In sub-Saharan Africa alone, the household electrification rate in 2018 averaged 45 per cent, while the population without electricity reached 591 million. There is also a vast gap in stable electricity supply between urban rich and urban poor households. For the growing population of Africa dependent on carbon-emitting sources, a decarbonisation approach to facilitate global climate action is required.

Moreover, intra- and inter-city transportation is primarily by fossil-fuel-powered modes on land and water. Besides, the mass migration of people from rural to urban centres for a better life and healthcare services increases the use of a transport system that operates on fossil fuel. Meanwhile, many African cities are challenged by poor planning, absent or inefficient waste management, the burning of charcoal for cooking, and continued use of fossil fuels to power telecommunication sites, healthcare institutions, and other settlement-related areas. Furthermore, African economies are carbon-based, with small-scale industries that survive from their linkages with the extractive industries. For instance, Nigeria depends on crude oil extraction and sale as a primary income earner. The fortunes of cities and communities and commercial activities in them are invariably tied to the hydrocarbon sector.

In developing low-carbon strategies for African cities, countries in the continent need to determine suitable pathways to decarbonisation that will allow for sustainable growth and development in line with national capabilities. Alongside the policy actions for decarbonisation, a strategic commitment to institution-building is the single most important intervention by any government wishing to lead the global response to climate change (Victor et al. 2019). A balanced and pragmatic decarbonisation process, through policies and regulatory instruments, is essential to enhance timely investment in cleaner, smarter, and more efficient energy technologies in the renewable energy industry context and for their application in cities. This will be through climate-smart systems for buildings and eco-friendly transport systems, especially in cities. Similarly, waste management and other green interventions are necessary for both decarbonisation and achieving the 2063 sustainable development agenda in Africa (African Union 2015). The IPCC (2018) observed that reducing worldwide emissions requires not just one low-carbon transition, but many. It requires rapid and far-reaching transitions in energy, land, urban infrastructure (including transport and buildings), and industrial systems. These system transitions are unprecedented in terms of scale, but not necessarily in terms of speed, and imply deep emissions reductions in all sectors.

So, this book evaluates low-carbon pathways for Africa by examining the unique contextual settings of cities within the continent. It adopts a reader-friendly style for the benefit of policymakers and other stakeholders in Africa and a global audience interested in the legal, sectoral, and governance dynamics of decarbonisation in African cities to achieve climate resilience. It is a book which examines the specific context of decarbonisation in sectors and cities within the continent, highlighting the frameworks for facilitating the process of decarbonisation. Unlike most other texts and works in the field, this book presents a multidimensional approach to carbon reduction in Africa by looking beyond the discourses in the energy sector to carbon reduction within sectors in cities. It highlights a contextual approach to policy formulation and implementation in the pursuit of low-carbon emission and energy transition which are essential in Africa's climate action.

1.2 Chapter Highlights

This book is structured into ten chapters covering six sectoral pathways—renewable energy, waste management, healthcare, telecommunication, education, and governance reconfigurations (for Petro-cities). Several cities across west, north, and east Africa are covered. Following this introduction (**Chap. 1**), the book seeks to examine the issues highlighted above by structuring the discussion around two broad parts:

I. Decarbonising African Cities: Strategies and Applications
II. Governance and Policy Approaches for Decarbonising African Cities

The first part of the book comprises **Chaps. 2–6**, while **Chaps. 7–9** make up the second part. A concluding chapter (**Chap. 10**) advances a low-carbon framework for African cities.

So far, Africa's rapid urbanisation has created poorly planned cities and the emergence of urban slums, with attendant problems. Moreover, the pursuit of decarbonising energy consumption through renewable sources has not been integrated into urban planning or city design in Africa. This fundamental issue is covered in **Chap. 2** by Akrofi, Okitasari, Ohunakin, and Azubuike, with the title "*Solar Urban Planning in African Cities: Challenges and Prospects*". Akrofi et al. introduce solar urban planning—a nascent concept with the potential for advancing urban development processes through design and/or planning. Given Africa's solar, wind, and tidal energy potentials, the chapter uniquely evaluates the prospects of incorporating solar urban planning into existing formalised systems. It applies the Preferred Reporting Items for Systematic Reviews and Meta-Analysis (PRISMA) framework and recommends the development of decision support systems and strategies for realising solar urban planning in Africa. Moreover, urban planning requires prioritisation of water, energy, and food (WEF) activities in order of importance (Liu et al. 2022) and in alignment with waste management.

Regarding the scale of human waste, agricultural waste, animal waste, air, land, and sea pollution around and within cities, Africa's carbon emission issues present real challenges for climate action. Interestingly, the job-creation prospects and macroeconomic potentials of recycling waste in many African cities are yet to be harnessed (Isaac et al. 2021). So, in **Chap. 3** ("*Contextualising Waste Management Operations Towards Low-Carbon African Cities*"), Muniafu and Nzembi propose contextualising waste management operations in Africa, asserting that governments prioritise waste management. The issues are examined in major African cities (Cape Town, Dar-es Salaam, Koshe, Lagos, Nairobi, and Ouagadougou) with linkages to health (SDG 3), access to clean water and sanitation (SDG 6). Furthermore, it highlights the possibilities of decarbonising African cities by focusing on recycling within the context of public-private partnerships. In conclusion, the authors posit that strong legislations should be enacted and implemented with good standards at national and municipal levels.

Linked to waste management and good hygiene are the outbreaks of epidemic and pandemic diseases (like Ebola and COVID-19) that have shown a considerable deficit in Africa's healthcare provisioning. Ironically, the healthcare sector that cares for patients—victims of poor waste management—also creates medical wastes that are hazardous to humans and the environment. This dilemma is considered, by Azubuike and Adeyemi in **Chap. 4**, amidst the burden of tropical diseases and pandemics in Nigeria. Entitled "*Innovative Strategies for Decarbonising the Healthcare Sector in Nigerian Cities*", the chapter highlights the general environmental impact of the healthcare sector in Africa. It is framed on ecological modernisation theory and subsequently applies the theory to discuss relevant literature. The chapter argues that urban healthcare facilities could apply the Top-Runner approach and climate-smart strategies to reduce their carbon footprint.

Africa's connection to the world via modern information and communications technology (ICT) is growing rapidly (Ihayere et al. 2020). This growth has heightened the extent of the continent's developmental needs. To sustain the socio-economic development of African cities, potentials for climate resilience in the operations of information and communications technology firms need to be explored (Gershon and Agbene 2021). So, Okundamiya and Wara explore decarbonisation pathways for the telecommunication sector in **Chap. 5** ("*Optimising Hybrid Power Systems for Sustainable Operation of Remote Telecommunication Infrastructure*"). Using a case study with 22-year meteorological datasets, the authors propose that telecommunications firms switch from fossil fuel-fired to carbon-neutral energy systems in their cell sites. By employing the energy-equilibrium procedures to optimise the model, it is shown that a wind-photovoltaic system is the optimal hybrid option for decarbonising telecommunication activities based on the case study. Focusing on the educational sector, and also using a case study, Wara and Okundamiya analysed the low-carbon choices for sustainably generating electricity in a University campus within Nigeria. As such, **Chap. 6** (“*Performance Analysis of a Grid-Linked Microgrid System in a University Campus*”) shows that a combination of grid-linked photovoltaic system and fuel cell microgrid system offers the possibility of reducing GHG emissions from most university campuses in cities within Nigeria. It is notable that most public and private Universities in Nigerian cities use a combination of intermittent public power supply and grid-diesel generator systems with high GHG emissions.

Muniafu, Ombewa, and Nzembi then redirect the focus to Nairobi by considering the GHG emissions reduction endeavours of the Kenyan Government, private sector, and civil society organisations. Hence, **Chap. 7** (“*Powering Action Towards Energising African Cities Sustainably: Perspectives from Kenya*”) captures different climate change mitigation actions that cut across cities—including strides in energy diversification aimed at reducing dependence on fossil fuels. Meanwhile, the sixth chapter also highlights a unique issue inherent in reducing carbon emissions within African urban cities—sustaining economic growth with rapidly increasing population, while also addressing energy poverty amidst worsening income inequalities (Adeleye et al. 2020). The situation is further complicated by the unreliability of electricity sector reforms and lack of competitive electricity markets in many developing countries (Ugwoke et al. 2020; Heffron et al. 2022). Amongst the notable propositions in the chapter is the need for Kenyan (and other African) cities to develop a strategy for recycling clean energy technology waste—like solar panels. It reinforces the discourse in Chap. 2 and proposes a job-creating pathway for addressing the environmental consequences of mass production and increasing utilisation of clean energy appliances.

Political-economic considerations, policy, and governance are at the core of effective implementation of carbon reduction strategies and climate action so they are covered in the last two chapters of the book. **Chapter 8** (“*The political economy of decarbonising African Petro-cities: Governance reconfigurations for the future*”) deals with the political economy of decarbonising urban cities of oil-rich countries in Africa (called Petro-cities) and proposes governance reconfigurations for the future. Abraham-Dukuma, Aholu, Nyokabi, and Dioha use a mix of theoretical and content analysis on four major Petro-cities (Hassi Messaoud, Luanda, Port Harcourt, and Tripoli). The chapter discusses governance, political, economic, and institutional interplays relevant to the decarbonisation processes of Petro-cities. Drawing from both national and city-scale analyses of the major challenges, the chapter proposes options for carbon emissions reduction, viz: optimising energy culture, promoting municipal energy democracy, and energy decentralisation, amongst others. It asserts that such regulatory, policy, and institutional governance frameworks set the stage to achieve a low-carbon economy. Alongside the policy actions for decarbonisation, a strategic commitment to institution-building is another important intervention needed for sustainably mitigating climate change (Victor et al. 2019). This dimension is critical to the decarbonisation agenda to avoid inconsistent and inadequate regulations and institutions in Africa.

In **Chap. 9** ("*To opt-in or to cop out: COP26 and the Policy Dynamics of Decarbonising African Cities*"), it is observed that while discussions at COP26 (the 26th Conference of Parties to the UN Framework Convention on Climate Change (UNFCCC) in Glasgow) have kept alive the hope of mitigating climate change, the real challenge is for African delegations to translate some of the agreements to specific policy and change instruments in their home countries. Meanwhile, African cities require substantial new infrastructure to meet nationally determined contributions (NDCs). Due to the failure of developed countries to fulfil pledges (of COP21), developing countries are reluctant to reduce carbon-emitting activities to levels considered inimical to their fragile, resource rent-dependent economies. In addition, decarbonisation will mean less dependence on fossil-fuel sources and reduced export revenue, and it will become more difficult for resource-rich African countries to secure resource-backed loans for development purposes. Accordingly, Asekomeh, Gershon, and Azubuike employ the Advocacy Coalition Framework (ACF) in the chapter to analyse how the beliefs and resources available to various coalitions within the policy subsystem can be construed to achieve policymaking that would operationalise the 'achievements' of the Glasgow Climate Pact to city decarbonisation measures for African cities. The review examines how the specific arrangements from membership of international alliances of cities and the mechanisms outlined in the African Adaptation Acceleration Program (AAAP) can be coalesced into a common set of interests to satisfy relevant stakeholders. The coordination observed in alliances of cities and the AAAP indicate that while external perturbations (like the climate emergency) would mean that coalitions are willing to reconsider their policy core beliefs, policy-oriented learning and negotiations to support policy change by considering secondary beliefs of coalitions will be required to agree with specific policy instruments to achieve decarbonisation.

1.3 Summary

Regulatory and policy instruments exist for low-carbon pathways in many countries, albeit, the strategies to achieve the underlying macroeconomic objectives are still lacking—especially, in Africa. This book addresses questions that policy-makers, investors, industry operators, lawyers, academics, and students engaged in the decarbonisation discourse seek answers to, including:

i. What determines the decarbonisation process, and how can the low-carbon process be implemented sustainably in Africa?
ii. How can Africa's cities be decarbonised to align with the climate action plan?
iii. How could investors in the energy sector respond to the peculiarities of African cities?
iv. What are the incentives needed to stimulate investment in the decarbonisation of Africa's cities (like Petro-cities)?
v. What are the governance, economic, and policy reconfigurations that could support timely investments towards reducing carbon emissions in Africa?

References

Adeleye, B.N., Gershon, O., Ogundipe, A., Owolabi, O., Ogunrinola, I., Adediran, O. (2020). Comparative investigation of the growth-poverty-inequality trilemma in Sub-Saharan Africa and Latin American and Caribbean Countries. *Heliyon, 6(12)*, e05631. *https://doi.org/10.1016/j.heliyon.2020.e05631*

African Union. (2015). Agenda 2063: The Africa we want. African Union Commission, September 2015. Retrived from https://au.int/sites/default/files/documents/36204-doc-agenda2063_popular_version_en.pdf.

Altieri, K., Trollip, H., Caetano, T., Hughes, A., Merven, B., Winkle, H. (2015). Pathways to deep decarbonisation in South Africa. http://deepdecarbonization.org/wp-content/uploads/2015/09/DDPP_ZAF.pdf

Daramola, P., Obindah, Gershon., Matthew, O., Ihayere, O., and Ejemeyovwi, J. (2021). Carbon Emission and Population Growth: Evidence from the Magna Cum Laude Oil Producing African Countries. IOP Conf. Series: Earth and Environmental Science 665 012038

Echeverri, L. G. (2018). Investing for rapid decarbonisation in cities. *Current Opinion in Environmental Sustainability*, 30, 42-51.

Fankhauser, S. and Jotzo, F. (2018). Economic growth and development with low-carbon energy. *Wiley Interdisciplinary Reviews: Climate Change*, 9(1), e495.

Fay, M., Hallegatte, S., Vogt-Schilb, A., Rozenberg, J., Narloch, U., & Kerr, T. (2015). *Decarbonizing Development: Three steps to a zero-carbon future*. The World Bank.

Gershon, O. and Asaolu, K. (2020). Evaporative quality of Nigeria's gasoline: truck loading perspective. *Energ. Ecol. and Environ.* https://doi.org/10.1007/s40974-020-00184-0

Gershon, O. and Agbene E. (2021). Adopting Hybrid Energy Technology for Carbon Emissions Reduction in Nigeria's Telecommunications Industry. IOP Conf. Ser.: Earth Environ. Sci. 665 012005. DOI:https://doi.org/10.1088/1755-1315/655/1/012048

Hafner, M., Tagliapietra, S., & de Strasser, L. (2018). The challenge of energy access in Africa. In *Energy in Africa* (pp. 1-21). Springer, Cham.

Heffron, R. J., Korner, M., Sumarno, T., Wagner, J., Weibelzahl, M., Fridgen, G., (2022). How different electricity pricing systems affect the energy trilemma: Assessing Indonesia's electricity market transition. *Energy Economics*. 107, 105663. https://doi.org/10.1016/j.eneco.2021.105663

Ihayere, O.B., Alege, P.O., Gershon, O., Ejemeyovwi, J.O., Daramola, P. (2020). Information communication technology access and use towards energy consumption in selected Sub Saharan Africa. International Journal of Energy Economics and Policy, 11(1), pp. 471–477. https://doi.org/10.32479/ijeep.9990

International Energy Agency (2019). SDG7: Data and Projections: Access to affordable, reliable, sustainable and modern energy for all. https://www.iea.org/reports/sdg7-data-and-projections/access-to-electricity

IPCC. (2018) Global warming of 1.5 0C; Summary for Policymakers, Intergovernmental Panel on Climate Change. https://www.ipcc.ch/sr15/chapter/spm/

Isaac, J. O., Olurinola, I. O., Gershon, O., Aderounmu, B. (2021). Working Conditions and Career Aspirations of Waste Pickers in Lagos State. Recycling, 6(1), 1; https://doi.org/10.3390/recycling6010001

Liu, S. K., Lin, Z. E., and Chiueh, P. T. (2022). Improving urban sustainability and resilience with the optimal arrangement of water-energy-food related practices. *Science of the Total Environment*. 812. DOI: https://doi.org/10.1016/j.scitotenv.2021.152559

Nalule, V. and Azubuike, S. I. (2020). Challenges and Opportunities for Energy Transition and Decarbonisation in Southern African Countries in Oyewunmi, T., Crossley, P., Talus, K and Sourgens, F (eds) *Decarbonisation and the Energy*

Industry: Law, Policy and Regulation in Low-Carbon Energy Markets. Bloomsbury Publishing. https://doi.org/10.5040/9781509932931.ch-016

NEA (2019), *The Costs of Decarbonisation: System Costs with High Shares of Nuclear and Renewables,* OECD Publishing, Paris, https://doi.org/10.1787/9789264312180-en

Ugwoke, B., Gershon, O., Becchio, C., Corgnati, S.P., and Leone, P. (2020). A review of Nigerian energy access studies: The story told so far. Renewable and Sustainable Energy Reviews. 120, 109646

Victor, D.G., Geels, F.W. and Sharpe, S. (2019) Accelerating the Low Carbon Transition: The Case for Stronger, More Targeted and Coordinated International Action. Brookings http://www.energy-transitions.org/sites/default/files/Accelerating-The-Transitions_Report.pdf

PART I

Decarbonising African Cities: Strategies and Applications

CHAPTER 2

Solar Urban Planning in African Cities: Challenges and Prospects

Mark M. Akrofi, Mahesti Okitasari, Olayinka S. Ohunakin, and Smith I. Azubuike

Abstract This chapter aims to identify and evaluate the opportunities and challenges of solar urban planning in Africa's cities. Using the Preferred Reporting Items for Systematic Reviews and Meta-Analysis (PRISMA) guidelines, 37 publications were selected and thematically analysed. It

M. M. Akrofi (✉) • M. Okitasari
Institute for the Advanced Study of Sustainability, United Nations University, Tokyo, Japan
e-mail: akrofi@student.unu.edu

O. S. Ohunakin
The Energy and Environment Research Group (TEERG), Mechanical Engineering Department, Covenant University, Ota, Nigeria

Faculty of Engineering & the Built Environment, University of Johannesburg, Johannesburg, South Africa

S. I. Azubuike
Durham Law School, Durham University, Durham, UK

S. I. Azubuike et al. (eds.), *Decarbonisation Pathways for African Cities*, Palgrave Studies in Climate Resilient Societies,
https://doi.org/10.1007/978-3-031-14006-8_2

emerged that the vast solar energy potentials, formalised urban planning systems, declining costs, and rising acceptance of solar PVs are key opportunities for solar urban planning. Urban informality, inadequate technical expertise, regulatory bottlenecks, and non-compliance with building regulations are critical barriers to Africa's solar urban planning. Research on the subject was found to be limited. Studies that explore decision support systems and strategies for integrating solar concerns into urban planning using multi-criteria assessments will be instrumental for realising this concept in Africa.

Keywords Africa • Cities • Energy transition • Solar energy • Sustainability • Urban planning

2.1 Introduction

2.1.1 Context

Africa has experienced immense economic growth over the past two decades, and it is expected to witness more growth over the coming decades. The International Monetary Fund (IMF 2020) projects that the continent's Gross Domestic Product (GDP) growth will increase from 3.8% in 2020 to about 4.2% in 2024, making it the world's second-largest economy (on a continental basis) only after Asia. Much of this growth is propelled by fossil-based energy, mostly from cities that function as industrial and economic centres. The challenge of decarbonising economic growth is compounded by the need to meet the energy needs of a fast-growing population, with significant gaps in current energy supply and demand. An estimated 95 million urban dwellers in Africa do not have access to electricity (World Bank 2017). More than half of the continent's countries have less than 80% electricity access in their urban areas (see Fig. 2.1).

Recent advances in renewable energy applications, especially solar photovoltaics (PV), provide an enormous opportunity to advance decarbonisation and clean energy access in African cities. An effective strategy to pursue this goal is through the integration of solar PVs on buildings. According to the International Energy Agency (IEA 2014), by integrating solar PVs into buildings, more than half of the global solar capacity could

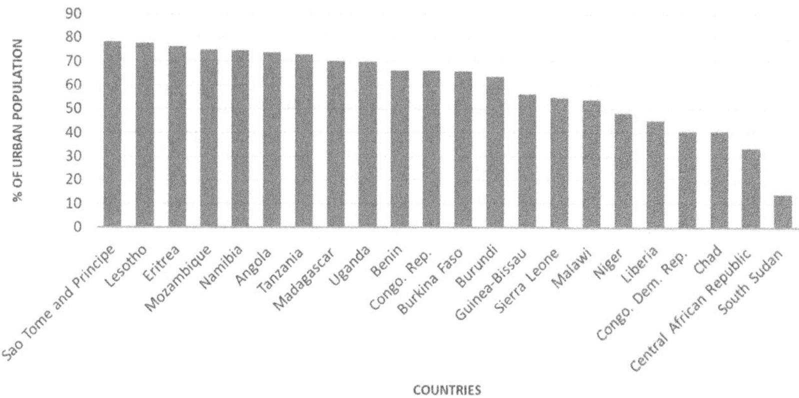

Fig. 2.1 Electricity access in urban areas as a percentage of the urban population (blue), and the urban population as a percentage of the total population (red) for sub-Saharan African countries with less than 80% electricity access in urban areas (2014). Source: IRENA (2019)

be utilised by 2050. This integration is crucial in cities, which currently consume about 78% of the energy produced globally, and account for more than 60% of greenhouse gas emissions (United Nations 2018). However, in many instances, considerations for Building Integrated Solar Photovoltaics (BIPV) are not made in the early stages of the urban design/ planning process.

Consequently, such integrations often lead to suboptimal and unattractive outcomes, which fail to stimulate the uptake of these technologies in cities (Kanters et al. 2013). To overcome this challenge, scholars now widely propose that solar energy concerns need to be integrated early in the urban planning process, giving rise to the term *solar urban planning* (Haine and Blumberga 2016; Lobaccaro et al. 2017b). Implementing solar energy in urban planning deals with the connections between solar energy and urban morphology, land use, and spatial structure of cities (Amado and Poggi 2014), regulations, and socio-demographic factors (Kanters and Wall 2018; Wall et al. 2017). Unearthing and understanding these barriers are vital for the widespread diffusion and utilisation of solar PVs in cities. However, much of the existing research on the subject has focused on Europe and other developed regions (Akrofi and Okitasari

2022). As a result, little is known about the prospects of solar urban planning in Africa.

Even though there are a few studies on decentralised generation and solar home systems (SHSs) in Africa (e.g. Barau et al. 2020; Boamah and Rothfuß 2018; Haine and Blumberga 2016), studies that deal with the actual implementation of solar energy in the urban planning process are limited. This chapter examines the prospects and challenges of solar urban planning in Africa based on a review of the available literature. As a first step, this review aims to outline key thematic issues regarding opportunities and challenges for solar urban planning in Africa and provides a baseline for more empirical studies/analyses, enhancing this concept's practicality in African cities. This chapter addresses the critical question: What are the prospects and challenges of solar urban planning in African cities?

2.1.2 Conceptualising Solar Urban Planning

Hanna (2016) views solar urban planning as "...a deeper dive into the notion that urban planning practices can specifically include, advance, and fortify the incorporation of solar and ultimately rely upon it as a substantial tool for addressing global warming" (p.4). Likewise, Amado and Poggi (2014) describe solar urban planning as an approach that utilises solar energy as an urban design principle to enhance energy efficiency and sustainable energy access in pre-existing and new urban developments. Thus, solar urban planning denotes a novel approach where solar energy development in urban areas and urban planning processes are no longer considered two independent activities but rather an integrated process that yields sustainable outcomes. Amado and Poggi (2012) proposed a framework to depict this process in Fig. 2.2.

The framework can be categorised into three main aspects—governance, socio-economic, and technological. The intervention programme definition, which entails goal-setting, may be seen as a governance/political process where renewable energy and carbon emission targets to be achieved through solar urban planning are set. The site analysis part of the framework could be described as the socio-economic phase since it primarily deals with social, economic, and environmental parameters such as demography, electricity prices, and climatic factors. The solar analysis aspect of this socio-economic group will best fit into the technical/technological category, which mainly involves stages four and five in the framework. These are the steps where simulations, master plans, and architectural designs are

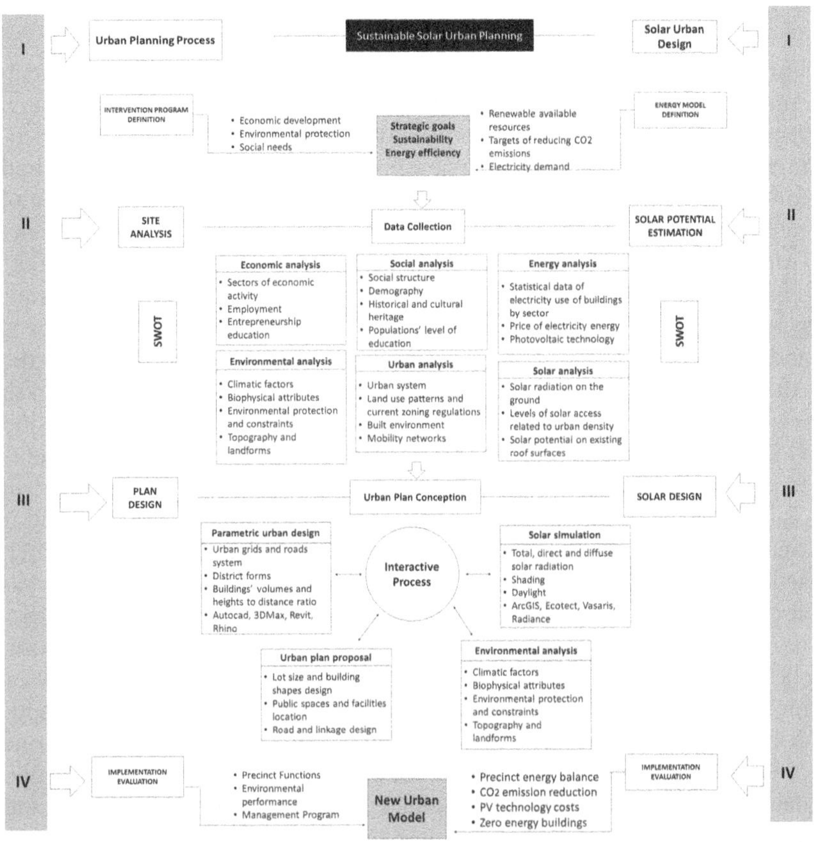

Fig. 2.2 Solar urban planning process. Source: Amado and Poggi (2012)

made and implemented. Key opportunities and challenges associated with these stages of solar urban planning constitute the focus of this chapter. Existing literature is gleaned to identify prospects and challenges associated with the process in the African context. In the ensuing section, the method adopted for the review is explained, while results, conclusions, and recommendations are discussed in Sects. 2.3 and 2.4, respectively.

2.2 Methods

Using the Preferred Reporting Items for Systematic Reviews and Meta-Analysis (PRISMA) guidelines (Moher et al. 2009), a well-defined search criterion was established to identify and select relevant publications/documents for the review. PRISMA is a well-known and widely applied method for systematic reviews. The method's transparency, reliability, and ease of replication helps to avoid biases and poor reporting in systematic reviews (Moher et al. 2009). Hence, providing a robust platform for this study. The criteria used for document search included the document types: original research articles, reviews, conference papers, book chapters, reports, and policy documents on solar energy and urban planning relevant to the African context. Documents authored in languages other than English were excluded. The SCOPUS database, ScienceDirect database, and Google search engine were used to identify and select relevant documents for inclusion. The search string ("solar energy" OR "BIPV") AND "urban planning" AND "Africa" was used in the ScienceDirect database, and it yielded 337 documents. The document type was then limited to research articles, review articles, books, and book chapters. The time period was also limited to articles published within the past ten years. These restrictions reduced the number of documents to 259. In the Scopus database, the search string TITLE-ABS-KEY ("solar energy" AND "urban planning" AND "Africa") was used, resulting in 5 research articles. The Scopus documents were merged with the ScienceDirect documents, and one duplicate was removed. Hence, a total of 263 documents from these two databases were screened for eligibility. After screening the titles and abstracts, 52 documents were selected. A further screening of the full texts of these 52 articles led to the exclusion of 23 articles; hence, a total of 29 articles were selected for inclusion.

Grey literature such as reports and working papers from key organisations such as IRENA, the UN, and the World Bank were identified through the Google search engine using the search terms; solar energy + urban planning + Africa. In all, 37 documents were selected. All selected documents were carefully read to identify prospects and challenges related to solar urban planning. The results were categorised and discussed in themes. Figure 2.3 gives an overview of the selection process.

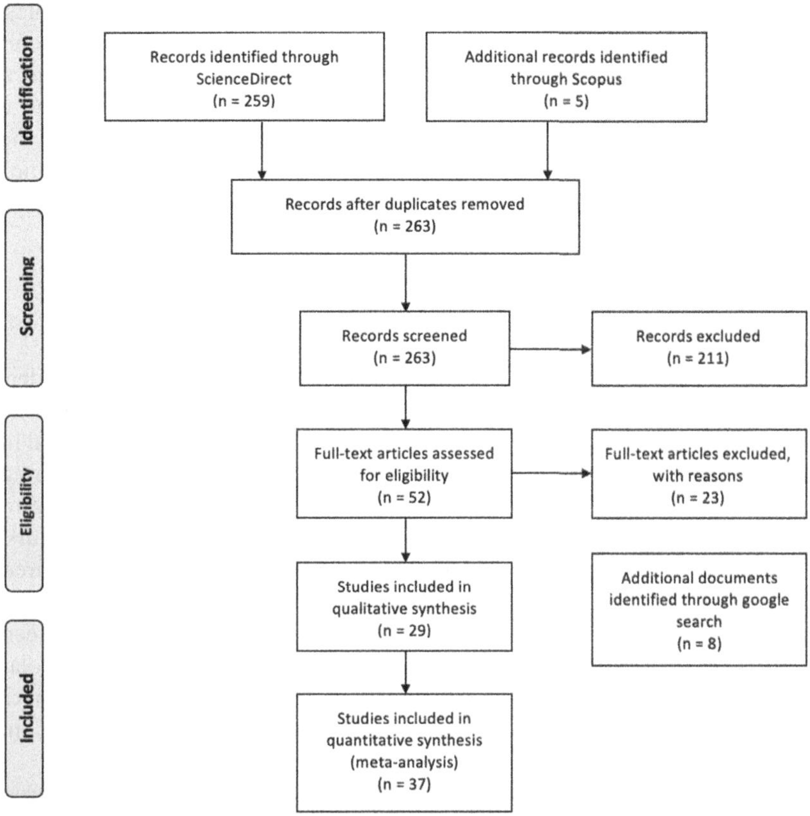

Fig. 2.3 Document identification and selection process. Adapted from Page et al. (2021)

2.3 Results and Discussion

2.3.1 Characteristics of the Selected Documents

Majority of the selected documents were journal articles. The trend of publications on the subject saw a sharp rise from 2015 and has since increased. This increment reflects the calls for clean energy transitions following the adoption of the Sustainable Development Goals (SDGs) and the Paris Agreement in 2015. Thus, the rising trends in publications on solar urban planning in Africa are consistent with these global efforts, which form a core part of the 2030 Agenda for Sustainable Development. Of the 37 documents selected, 51% had a geographic focus in at least a country in North, South, East, or West Africa. In contrast, 49% had a broad focus on the entire continent or multiple countries across the regional subdivisions. Specifically, the documents came from eight countries, and their respective proportions are shown in Fig. 2.4.

2.3.2 *Prospects of Solar Urban Planning in Africa*

2.3.2.1 Huge Solar Energy Potentials in Cities

A solar potential analysis is the foremost and most important consideration in the solar urban planning process. This analysis is essential because if adequate power cannot be produced from the available solar potential in a particular locality to meet its energy needs, the solar urban planning process will lead to unsatisfactory outcomes (Lauka et al. 2018). Due to its location in the tropics, Africa receives a high number of sunshine hours. The solar irradiance in African cities, in particular, is very high compared to their European counterparts. According to the IRENA (2016), the average solar irradiation in African countries' capital cities ranges between 1750 and 2500 kWh/m^2/year, with 39 of the 54 countries having a solar potential that exceeds 2000 kWh/m^2/year (IRENA 2016). Studies have also found the rooftop solar potential to be enormous and viable in African cities.

Mukisa et al. (2019) found that a potential energy yield of between 1046 kWh/kW–1344 kWh/kW could be generated for all roof orientations of industrial buildings in Uganda. Lauka et al. (2018) also discovered that installing solar PVs on roofs and facades can generate 2861 MWh/y a year (77.3% of the annual per capita consumption) and 6922 MWh/m^2/year for existing buildings and new urban areas, respectively in Ibenbadis, Algeria by 2025. In South Africa, Okunlola et al. (2019) noted that economically viable residential rooftop solar PV potential amounts to about 11.2 GW in metropolitan municipalities alone. Data from the World Bank's Global Solar Atlas (2020) further shows that annual solar PV power output for small residential and commercial systems in most African cities is more than 1 megawatt hours (MWh) and 100 MWh, respectively (see Fig. 2.5). These potentials present enormous

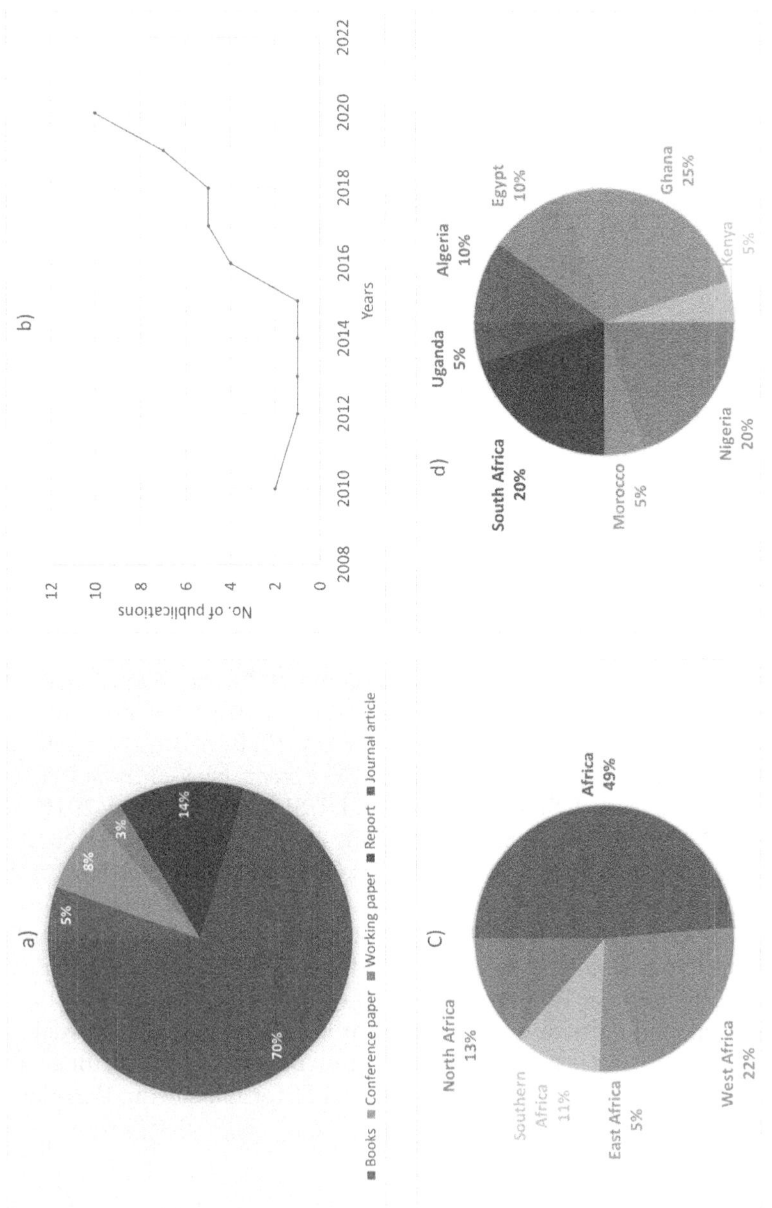

Fig. 2.4 Characteristics of selected documents: types of documents (**a**), trend of publications (**b**), regional focus of publications (**c**), and the number of publications by country (**d**)

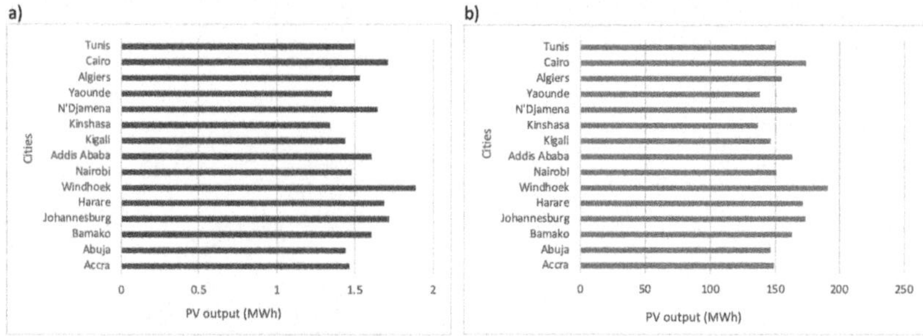

Fig. 2.5 Annual solar PV energy yield for small residential (**a**) and medium commercial (**b**) PV systems in selected African cities. Source: Authors' construct based on data from the World Bank (2020)

opportunities for solar energy integration in African cities, which can be facilitated through solar urban planning.

2.3.2.2 Declining Costs of Solar PV Technologies

The high cost of solar PV technologies has been one of the main challenges to their deployment and utilisation in Africa. However, the price of solar PV has reduced substantially in the last decade. Globally, solar PV costs have dropped by 82% between 2010 and 2020, the sharpest decline among other forms of renewable energies such as offshore wind, onshore wind, and concentrated solar power (IRENA 2020). Also, the levelised cost of electricity (LCOE) of residential PV systems by country and market decreased from USD0.301/kWh and USD0.455/kWh in 2010 to between USD0.063/kWh and USD0.265/kWh in 2019 (IRENA 2020). Figure 2.6 presents the trends in solar PV power costs from 2010 to 2019.

According to the IRENA (2016), the cost of utility-scale solar PV projects has decreased 61% since 2012, while residential solar systems satisfy the annual household energy needs for as low as USD56 per year in Africa. These declining solar PV costs provide an enormous opportunity for their deployment and utilisation in Africa since they can increase the affordability of solar PV systems. Several studies (e.g. Barau et al. 2020; Boamah 2020) have identified affordability as a vital factor influencing the adoption of solar PV systems in African cities.

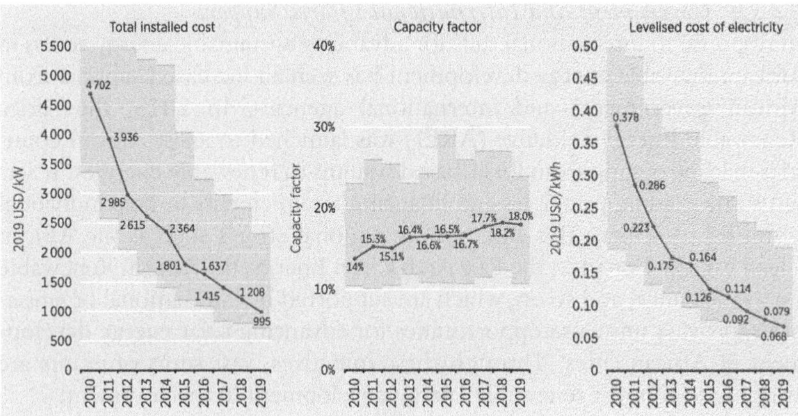

Fig. 2.6 Global weighted average total installed costs, capacity factors, and LCOE for solar PV, 2010–2019. Source: IRENA (2020)

2.3.2.3 Formalised Urban Planning Processes

Sikder et al. (2016) proposed that the existing formal urban planning process needs to be fused with energy planning to achieve energy-optimised urban planning. They argued that considerations for energy integration should be at the early design stage of residential urban planning, noting that solar PV offer significant opportunities for such integration (Sikder et al. 2016). Fortunately, most African countries already have formal urban planning processes and regulations, which provide the baseline upon which solar urban planning can be advanced. There are procedures for zoning, land use, master planning, and building regulations to be followed for orderly urban development in most African countries (Silva 2015).

These procedures are implemented mainly through decentralised departments such as town and country planning units at the municipal level. These units oversee building regulations and issue building permits for various land-use developments. For example, a three-tier planning system is practised in Ghana, where local and structural plans are prepared and operationalised based on a national spatial development framework. This system is implemented through a decentralised Town and Country Planning Department, which operates at the national, regional, and local government levels. Similar spatial planning and development control approaches are observed in Kenya, Ethiopia, Zimbabwe, Egypt, and South Africa, amongst others.

2.3.2.4 *Government and International Efforts/Support*

In response to international calls for advancing sustainable energy access in Africa, renewable energy development has seen an increased support from African governments and international agencies. In 2015, the Africa Renewable Energy Initiative (AREI) was launched to assist African countries in leapfrogging from fossil-based systems to renewable energies. It sets out some guiding principles and financing arrangements to help individual countries to achieve this aim. Similar regional efforts such as the African Clean Energy Corridor, the Pan-Arab Clean Energy, the African Renewable Energy Alliance, and so on, which are supported by international organisations, provide immense opportunities for advancing solar energy development in African cities. Through these initiatives, vast sums of grants are available to support renewable energy development in the continent.

Notwithstanding these regional commitments, African governments have taken several local initiatives to promote the use of solar energy in cities. For instance, in Ghana, the government launched a national rooftop solar program in 2015. The program sought to install 20,000 solar home systems through a capital subsidy program to achieve a 200 MW peak load relief on the national grid. Similarly, Zimbabwe's government launched a 20 MW off-grid solar PV rooftop program in 2017, targeting small and medium-sized enterprises (SMEs) that rely on diesel generators. The International Energy Agency (IEA 2020) notes that governments' relief packages in response to COVID-19 have prioritised investments in renewable energy systems to support off-grid electrification and health facilities. Such government support, not only in terms of funding but also favourable regulations, is essential for formalising solar energy considerations into urban planning practices.

2.3.2.5 *Rising Awareness and User Acceptance of Solar PV Applications*

Limited awareness and acceptance of renewable energy applications have been identified as constraints to deploying and utilising solar PV in Africa (Adenle 2020). However, this status quo is rapidly changing, especially in urban areas, due to the rising needs for reliable power supply and energy independence. Boamah (2020) recounts that frequent power outages and a show of social status are compelling many urban dwellers in Ghana to install solar home systems. Other studies have shown a high level of awareness and approval of the development of renewable energy technologies (RETs) in many African countries (Nwokocha et al. 2018; Oluoch et al.

2020; Strazzera and Statzu 2017). While user acceptance is increasing, there is also a strong political acceptance, which is evident in the various political commitments being made to support solar PV utilisation in Africa, as outlined in Sect. 2.3.2.4. Both user and political acceptance are vital for the success of solar urban planning in Africa. Hence, these rising trends provide a good opportunity for building low-carbon cities by integrating solar considerations into urban planning processes in Africa.

2.3.3 Challenges of Solar Urban Planning in African Cities

2.3.3.1 Urban Morphology and Land Use Challenges

Urban form, which entails cities' configuration and land-use patterns, plays a significant role in maximising solar energy potentials (Lobaccaro et al. 2017a). For example, Lobacarro et al. (2017a) found that when elements of urban morphology such as the height of buildings and the distance between them are optimised, the solar energy potential in the city of Trondheim could be increased by 25%. The morphology of many African cities will pose a challenge to solar urban planning, especially in existing urban areas. Barau et al. (2020) opined that sub-Saharan African urban households might face limited choices in their energy transitions due to urban informality. Such informalities pose a significant challenge to solar urban planning in existing urban areas. Existing buildings, for instance, may be unable to accommodate the additional weight that rooftop solar PVs would exert. The multi-occupancy nature of urban housing in many African cities could also present challenges to installing and managing solar home systems, especially rooftop solar PV. Where such systems cannot meet all occupants' needs, usage and maintenance issues could be problematic.

Aside from urban morphology's influence on the solar potential in cities, it also plays a significant role in solar urban planning from the perspective of what Lundgren and Dahlberg (2018) termed "energy landscape." Due to their proximity to places of abode, large-scale solar arrays, in particular, transform the urban landscape into a spatial and cultural medium that shapes the perception of solar energy generation (Lundgren and Dahlberg 2018). Hanna (2016) contends that cities epitomise our cultural lineage; hence, solar energy solutions must be integrated into the urban environment using urban design principles that incorporate pre-existing urban contexts. This raises the challenge of maintaining the cultural value of buildings while integrating solar PV. Cities in North African countries such as Egypt, Algeria, Morocco, and Tunisia are easily identified by their architectural heritage visible in their urban form. Formalising solar energy into urban planning procedures for such cities will need to be consistent with their architectural heritage. For instance, Lundgren and Dahlberg (2018) note that where the integration of solar energy in buildings in such landscapes meets social acceptance, it also translates into an acceptance of a modified landscape.

2.3.3.2 Techno-economic Challenges

Even though the cost of solar PV technology has declined rapidly over the past years, these cost reductions primarily benefit utility-scale solar PV projects (IRENA 2016). The cost of solar home systems remains comparatively high. Amidst the cost decline in solar PVs, the prices of storage batteries and inverters have not been significantly reduced. Those with higher capacities that can meet most user needs remain expensive and affordable by only wealthier households. For instance, Ghana's rooftop solar PV program favours wealthy urban households who could afford the needed Balance of System (BOS) components and have prior experience using solar home systems (Boamah and Rothfuß 2018). This cost factor may deepen inequalities if appropriate measures are not implemented to ensure equity and equality in the solar urban planning process.

The availability of reliable data and technical expertise is an essential prerequisite for solar urban planning. However, most of these data, especially LiDAR data, are unavailable for many African countries. Google's Project Sunroof is another useful data source for rooftop solar potential analysis, yet at present, it does not cover the African region. Besides these data requirements, technical and economic analysis regarding architectural integration issues, solar zoning, and so on, require technical expertise and specialised software such as PVsyst, SolarGIS, Aurora Solar, and so on. The inadequacy of such expertise in Africa is a significant challenge for solar urban planning. Such expertise is required on the part of urban planners and real estate developers, but some basic knowledge about handling solar home systems is also needed by the users (Barau et al. 2020; Boamah 2020).

2.3.3.3 Legal and Regulatory Challenges

Compliance with building regulations and standards remains a major challenge in African cities. Where they exist, building energy regulations are largely ignored (Iwaro and Mwasha 2010). Iwaro and Mwasha (2010) found that 26 countries among the 60 developing countries that they surveyed did not have building energy regulations. Fifteen of these 26 countries were in Africa. Where solar urban planning regulations are in place, enforcement and compliance barriers will be significant hurdles to be resolved. Ownership of land and land resources also plays a key role in urban planning and presents a challenge to integrating solar energy in the urban environment (Lundgren and Dahlberg 2018). Most sub-Saharan African cities are characterised by informal land markets, where individuals and traditional rulers own the land. However, some countries' existing legislative frameworks are unclear and lack focus on customary land ownership and administration in urban planning.

For medium to large-scale solar PV arrays in urban areas, access to land is a critical element. Thus, urban planners may need to balance customary landowners' interests with solar energy concerns in the solar urban planning process. The urban development landscape in Africa is also characterised by multiple actors, ranging from government agencies to private sector developers and investors. In the solar urban planning process, the challenge of balancing overlapping roles, different interests, and unbalanced power relations is imminent (Nuhu 2019). Also, conflicts between utilities and off-grid companies are likely to occur since a move towards decentralised energy will give greater energy independence to end-users and possibly decrease their energy patronage from utilities. Such a situation will take a toll on the revenue of utilities. Given that utilities in many African countries are state-owned, a strong political commitment will be needed to enact regulations favouring distributed generation.

2.3.3.4 Environmental Challenges

Balancing solar energy concerns with environmental sustainability is essential in decarbonising cities and building sustainable societies. From the production stage, some of the materials used in manufacturing solar PV are toxic and risked being released into the environment and present adverse problems if care is not taken (Rabaia et al. 2021). During their installation, biodiversity and ecosystem disruptions are likely to occur, especially for ground-mounted solar PV in urban settings. In the case of rooftop solar PV, competition for space between solar PV and green roofs also presents a challenge. Green roofs are essential for biodiversity, lowering urban air temperature, and mitigating the urban heat island effect. These benefits imply that their simultaneous use of solar PV would significantly benefit the urban environment. However, competition for space arises when both PV and green roofs are to be developed on the same roof. When appropriately integrated, green roofs can boost solar PV functionality through dust particle absorption and cooling effects (Sattler et al. 2020). On the other hand, adverse effects such as the shading of PV panels by the plants may occur if the integration is not done correctly; hence solutions that proffer synergistic relationships must be developed (Sattler et al. 2020).

2.3.3.5 SWOT Analysis

There are several prospects for decarbonising African cities through solar urban planning, although significant challenges need to be addressed. Some of these prospects and challenges are external, while others are internal in the continent. Examples include formalised urban planning processes (internal) and declining solar PV costs (external). The strengths, weaknesses, opportunities, and threats (SWOT) analytic tool (Fig. 2.7) is applied in this section to categorise various internal and external factors and how strengths and opportunities can be leveraged to address weaknesses and threats.

Opportunities such as new software tools can help urban planners in Africa to conduct solar potential analysis, perform solar building simulations, and prepare solar-integrated urban plans. A review by Jakica (2018) shows that about 200 solar design tools have been developed, ranging from solar potential analysis tools to architectural solar integration tools. However, inadequate technical expertise is a weakness that needs to be resolved to make full use of these software tools. This weakness could be resolved through training programs for urban planners and mainstreaming

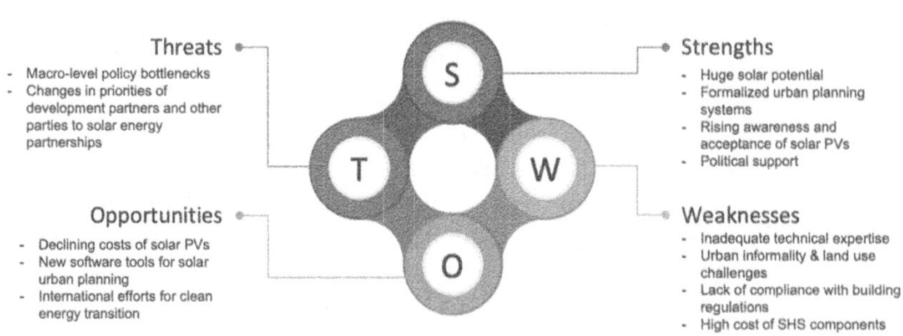

Fig. 2.7 SWOT analysis

new software tools and solar urban planning as a study area into the curriculum of urban planning departments of African universities. Teaching and learning manuals on solar urban planning for tertiary education were developed as part of the IEA Solar Heating and Cooling Program (TASK 51). These manuals are available and can be adapted and used for teaching and learning in African universities.

Another critical weakness is urban informality. While urban informality is a weakness to solar urban planning, this weakness is characteristic of existing urban areas. New urban developments in African cities provide an avenue to facilitate solar urban planning. Gated communities have become prominent in many African cities, especially in peri-urban areas, which are primarily driven by private real estate developers (see Fig. 2.8). This type of development could be leveraged through favourable regulations and policy support for real estate developers to integrate solar considerations into their building designs and urban environment.

International efforts for decarbonising and building sustainable cities through clean energy transitions are also on the rise. While these efforts serve as opportunities that could aid solar energy integration in cities, some macro-level policy conflicts could inhibit solar energy development at the local level. Hanna (2016) cites an example where the World Trade

Fig. 2.8 Aerial image of a gated community in Accra, Ghana (picture taken by the authors)

Organization (WTO) ruled against the Indian government's decision to rely solely on their domestic solar technologies for future projects. The government's decision was seen to conflict with the General Agreement on Tariffs and Trade (GATT) of the WTO. Such macro-level measures could be counterproductive and undermine the deployment of solar PV within countries (Hanna 2016). Changes in the priorities of international organisations and governments could also affect their commitment to partnerships that seek to boost decarbonisation efforts through solar energy development in cities. Weaknesses such as the high cost of solar home system (SHS) components could be resolved through government subsidy schemes. Enforcement of building standards and strict compliance checks can also help to address the weakness of non-compliance.

2.4 Conclusion

Cities function as important centres of economic growth. Yet, they contribute substantially to global carbon emissions due to the enormous amount of energy they consume. Decarbonising cities is, therefore, critical in addressing the climate crisis. Solar urban planning is a nascent strategy that can significantly facilitate the integration of solar PV in cities. In this chapter, we discussed the prospects and possible barriers to realising this concept in Africa. Based on a review of available literature, the emergent findings suggest more prospects than challenges for solar urban planning in Africa. The SWOT analysis shows there are more strengths and opportunities than weaknesses and threats. These strengths and opportunities can be leveraged to facilitate the integration of solar PV systems in African cities and consequently engender a transition to clean energy for decarbonised and sustainable cities. Empirical studies on solar urban planning are, however, limited in Africa.

While some studies on solar home systems and BIPVs exist, studies that deal with integrating solar energy considerations into traditional urban planning processes are lacking. Future studies dealing with techno-economic feasibility, decision support systems, and solar urban planning regulations will be crucial for realising this concept in Africa. Given the urban environment's multi-actor nature, studies exploring actor interests and best policy alternatives for formalising solar concerns into urban planning using multi-criteria assessments will be instrumental. Research on appropriate strategies for mainstreaming solar concerns into urban planning for pre-existing urban areas (urban renewal or regeneration plans) is essential for dealing with urban informality and morphological constraints to solar energy integration in African cities.

Acknowledgement This chapter was extracted from the first author's ongoing PhD thesis at the United Nations University Institute for the Advanced Study of Sustainability (UNU-IAS). We are grateful to the Japan Foundation for the United Nations University (JFUNU), which provided a scholarship for his doctoral studies at the UNU-IAS.

References

Adenle, A. A. (2020). Assessment of solar energy technologies in Africa-opportunities and challenges in meeting the 2030 agenda and sustainable development goals. *Energy Policy, 137*, 111180. https://doi.org/10.1016/j.enpol.2019.111180

Akrofi, M. M., & Okitasari, M. (2022). Integrating solar energy considerations into urban planning for low carbon cities: a systematic review of the state-of-the-art. Urban Governance. https://doi.org/10.1016/j.ugj.2022.04.002

Amado, M., & Poggi, F. (2014). Solar urban planning: a parametric approach. *Energy Procedia, 48*, 1539–1548. https://doi.org/10.1016/j.egypro.2014.02.174

Amado, M., & Poggi, F. (2012). Towards solar urban planning: A new step for better energy performance. *Energy Procedia, 30*, 1261–1273. https://doi.org/10.1016/j.egypro.2012.11.139

Barau, A. S., Abubakar, A. H., & Kiyawa, A.-H. I. (2020). Not There Yet: Mapping Inhibitions to Solar Energy Utilisation by Households in African Informal Urban Neighbourhoods. *Sustainability 2020, Vol. 12, Page 840*, *12*(3), 840. https://doi.org/10.3390/SU12030840

Boamah, F. (2020). Desirable or debatable? Putting Africa's decentralised solar energy futures in context. *Energy Research and Social Science, 62*, 101390. https://doi.org/10.1016/j.erss.2019.101390

Boamah, F., & Rothfuß, E. (2018). From technical innovations towards social practices and socio-technical transition? Re-thinking the transition to decentralised solar PV electrification in Africa. *Energy Research and Social Science, 42*, 1–10. https://doi.org/10.1016/j.erss.2018.02.019

Haine, K., & Blumberga, D. (2016). Towards Solar Urban Planning: A New Step for Better Energy Performance Case of Study Ibenbadis, Constantine (Algeria). *Energy Procedia, 95*, 145–152. https://doi.org/10.1016/j.egypro.2016.09.036

Hanna, T. (2016). Solar Urban Planning: Addressing barriers and conflicts specific to Renewable Energy Policy and the current field and practice of Urban

Planning within the context of a changing Climate. *Capstone Collection, 2956.* http://digitalcollections.sit.edu/capstones/2956

IEA. (2014). *Technology Roadmap Solar Photovoltaic Energy - 2014 edition.* www.iea.org

IEA. (2020). *Renewables 2020: Analysis and forecast to 2025.* https://www.iea.org/reports/renewables-2020/solar-pv

IMF. (2020). *World Economic Outlook (October 2020) - Real GDP growth.* https://www-imf-org.link.unu.edu/external/datamapper/NGDP_RPCH@WEO/OEMDC/ADVEC/WEOWORLD/AFQ

IRENA. (2016). *Solar PV in Africa: Costs and Markets.* /publications/2016/Sep/Solar-PV-in-Africa-Costs-and-Markets

IRENA. (2019). *Solar simulators: Application to developing cities.* www.irena.org

IRENA. (2020). Renewable Power Generation Costs in 2019. In */publications/2020/Jun/Renewable-Power-Costs-in-2019.* /publications/2020/Jun/Renewable-Power-Costs-in-2019

Iwaro, J., & Mwasha, A. (2010). A review of building energy regulation and policy for energy conservation in developing countries. *Energy Policy, 38*(12), 7744–7755. https://doi.org/10.1016/j.enpol.2010.08.027

Jakica, N. (2018). State-of-the-art review of solar design tools and methods for assessing daylighting and solar potential for building-integrated photovoltaics. In *Renewable and Sustainable Energy Reviews* (Vol. 81, pp. 1296–1328). https://doi.org/10.1016/j.rser.2017.05.080

Kanters, J., Dubois, M. C., & Wall, M. (2013). Architects design process in solar-integrated architecture in Sweden. *Architectural Science Review, 56*(2), 141–151. https://doi.org/10.1080/00038628.2012.681031

Kanters, J., & Wall, M. (2018). Experiences from the urban planning process of a solar neighbourhood in Malmö, Sweden. *Urban, Planning and Transport Research, 6*(1), 54–80. https://doi.org/10.1080/21650020.2018.1478323

Lauka, D., Haine, K., Gusca, J., & Blumberga, D. (2018). Solar energy integration in future urban plans of the South and Nordic cities. *Energy Procedia, 152,* 1127–1132. https://doi.org/https://doi.org/10.1016/j.egypro.2018.09.137

Lobaccaro, G., Carlucci, S., Croce, S., Paparella, R., & Finocchiaro, L. (2017a). Boosting solar accessibility and potential of urban districts in the Nordic climate: A case study in Trondheim. *Solar Energy, 149,* 347–369. https://doi.org/10.1016/j.solener.2017.04.015

Lobaccaro, G., Lindkvist, C., Wall, M., & Wyckmans, A. (2017b). *Illustrative Perspective of Solar Energy in Urban Planning.* https://doi.org/10.18777/ieashc-task51-2017-0002

Lundgren, M., & Dahlberg, J. (2018). *Approaches, Methods and Tools for Solar Energy in Urban Planning.* https://doi.org/10.18777/ieashc-task51-2018-0004

Moher, D., Liberati, A., Tetzlaff, J., & Altman, D. G. (2009). Preferred reporting items for systematic reviews and meta-analyses: The PRISMA statement. In *BMJ (Online)* (Vol. 339, Issue 7716, pp. 332–336). British Medical Journal Publishing Group. https://doi.org/10.1136/bmj.b2535

Mukisa, N., Zamora, R., & Lie, T. T. (2019). Feasibility assessment of grid-tied rooftop solar photovoltaic systems for industrial sector application in Uganda. *Sustainable Energy Technologies and Assessments, 32*, 83–91. https://doi.org/10.1016/j.seta.2019.02.001

Nuhu, S. (2019). Peri-Urban Land Governance in Developing Countries: Understanding the Role, Interaction and Power Relation Among Actors in Tanzania. *Urban Forum, 30*(1), 1–16. https://doi.org/10.1007/s12132-018-9339-2

Nwokocha, C. O., Okoro, U. K., & Usoh, C. I. (2018). Photovoltaics in Nigeria – Awareness, attitude and expected benefit based on a qualitative survey across regions. *Renewable Energy, 116*, 176–182. https://doi.org/10.1016/j.renene.2017.09.070

Okunlola, A., Jacobs, D., Ntuli, N., Fourie, R., Nagel, L., & Helgenberger, S. (2019). *Consumer savings through solar PV self-consumption in South Africa: Assessing the co-benefits of decarbonising the power sector.* https://doi.org/10.2312/iass.2019.007

Oluoch, S., Lal, P., Susaeta, A., & Vedwan, N. (2020). Assessment of public awareness, acceptance and attitudes towards renewable energy in Kenya. *Scientific African, 9*, e00512. https://doi.org/10.1016/j.sciaf.2020.e00512

Page, M. J., McKenzie, J. E., Bossuyt, P. M., Boutron, I., Hoffmann, T. C., Mulrow, C. D., Shamseer, L., Tetzlaff, J. M., Akl, E. A., Brennan, S. E., Chou, R., Glanville, J., Grimshaw, J. M., Hróbjartsson, A., Lalu, M. M., Li, T., Loder, E. W., Mayo-Wilson, E., McDonald, S., ... Moher, D. (2021). The PRISMA 2020 statement: an updated guideline for reporting systematic reviews. *BMJ, 372*, n71. https://doi.org/10.1136/bmj.n71

Rabaia, M. K. H., Abdelkareem, M. A., Sayed, E. T., Elsaid, K., Chae, K. J., Wilberforce, T., & Olabi, A. G. (2021). Environmental impacts of solar energy systems: A review. *Science of the Total Environment, 754*, 141989. https://doi.org/10.1016/j.scitotenv.2020.141989

Sattler, S., Zluwa, I., & Österreicher, D. (2020). The "PV Rooftop Garden": Providing Recreational Green Roofs and Renewable Energy as a Multifunctional System within One Surface Area. *Applied Sciences, 10*(5), 1791. https://doi.org/10.3390/app10051791

Sikder, S., Eanes, F., Asmelash, H., Kar, S., & Koetter, T. (2016). The Contribution of Energy-Optimized Urban Planning to Efficient Resource Use–A Case Study on Residential Settlement Development in Dhaka City, Bangladesh. *Sustainability, 8*(2), 119. https://doi.org/10.3390/su8020119

Silva, C. N. (2015). *Urban Planning in Sub-Saharan Africa: Colonial and Post-Colonial Planning Cultures* (C. N. Silva (ed.); 1st ed.). Routledge.
Strazzera, E., & Statzu, V. (2017). Fostering photovoltaic technologies in Mediterranean cities: Consumers' demand and social acceptance. *Renewable Energy*, *102*, 361–371. https://doi.org/10.1016/j.renene.2016.10.056
United Nations. (2018). *Cities and Pollution contribute to climate change*. United Nations. https://www.un.org/en/climatechange/cities-pollution.shtml
Wall, M., Snow, M., Dahlberg, J., Lundgren, M., Lindkvist, C., Lobaccaro, G., Siems, T., Simon, K., Cristina, M., & Probst, M. (2017, October 28). Urban Planning for Solar Energy-IEA SHC TASK 51. *ISES Solar World Congress 2017*. https://doi.org/10.18086/swc.2017.37.06
World Bank. (2017). *World Bank Open Data | Data*. https://data.worldbank.org/
World Bank. (2020, December). *Global Solar Atlas*. https://globalsolaratlas.info

CHAPTER 3

Contextualising Waste Management Operations Towards Low-Carbon African Cities

Laura Muniafu and Nzembi Mutiso

Abstract Waste management in cities is influenced by the rapid increase in population due to rural-urban migration and a growing middle class. This is an increasing challenge in many African countries leading to great environmental hazards. Many cities are yet to prioritise waste management and decarbonisation because public funds are geared towards the elimination of poverty, job creation, education, and health for citizens. Governments should protect the environment as a priority to protect its citizens' general health. This chapter focuses on recycling as a pathway to decarbonisation, available flexible strategies and financial incentives and the different waste management practices in African cities. The chapter

L. Muniafu (✉)
Strathmore Extractives Industry Centre, Strathmore University, Nairobi, Kenya
e-mail: lmuniafu@strathmore.edu

N. Mutiso
Extractives Baraza, Strathmore University, Nairobi, Kenya
e-mail: rmutiso@strathmore.edu

S. I. Azubuike et al. (eds.), *Decarbonisation Pathways for African Cities*, Palgrave Studies in Climate Resilient Societies,
https://doi.org/10.1007/978-3-031-14006-8_3

concludes that strong legislation should be implemented at the national and municipality levels. Public-private participation is key to improving waste management in developing countries.

Keywords Solid waste management • Waste recycling • Decarbonisation of waste • Sustainable development • Clean development mechanism

3.1 Introduction

The urban population in Africa is increasing by 3.5% per annum. In a report compiled by the United Nations Environment Programme (2018), *Africa Waste Management Outlook,* it was highlighted that most African countries face development challenges even though Africa is aspiring to be part of the global economic and development agenda. Population growth, rapid urbanisation, a growing middle class, changing consumption habits and production patterns and global waste trade trafficking brings about waste generation in African cities. Improper waste management has serious health and environmental consequences (UNEP 2018).

3.2 Waste Management Governance and Policies in Africa

Africa has developed continental policies and strategies to address waste management in the continent. Most African countries have also ratified international conventions showing their commitment to effective waste management.

3.2.1 *International Conventions*

A notable international instrument regarding waste management is the Bamako Convention on the Ban of the Import into Africa and the Control Transboundary Movement and the Management of Hazardous Wastes within Africa of 1991. This Convention obligates countries to ban the import of hazardous and radioactive wastes. Countries that are signatories have developed laws in relation to this convention. Some of the laws include Angola's Regulation on Waste Management, 2012, Benin's Law No. 98-030 of 12 February 1999, Burundi's Law No 1/010 of 30th June 2000 on the Code of Environment, Cameroon's Law No 96/12 of August 1996. The 1992 United Nations Framework Convention on Climate Change also includes waste management and greenhouse gas emissions (GHG) regulation. However, most African countries have ratified these conventions but have not domesticated them into their national laws.

3.2.2 Regional Policies

Within the regional context, the Southern African Development Community (SADC): Regional Indicative Strategic Development Plan (2020–2030) is worthy of note. This development plan provides a framework for SADC member states to commit to the promotion of sound environmental management through pollution control, waste management, and environmental education.

Most African countries are yet to adopt these conventions and policies and it is unclear whether they have been translated into action. The non-implementation of waste and environmental legislation by African governments has brought about a culture of impunity and weakened the effectiveness of waste management. Across Africa, there is weak enforcement of legislation. Courts of law play an effective role in enforcing legislation if they have the means to make binding decisions (UNEP 2018).

3.3 Solid Waste and Recycling Practices in Africa

With only a 4% recycling rate, opportunities to develop a "secondary resources economy" are still largely unexplored in Africa. Current waste management practices have resulted in waste being overlooked for the value that it can provide to local economies (Godfrey et al. 2019).

The informal sector plays a major active role in collecting and diverting reusable and recyclable waste in Africa. This has led to the creation of jobs (Godfrey et al. 2019). However, there are existing gaps in waste collection, disposal, reuse, recycling, and recovery which has led to the emergence of social and technological innovations in Africa. Examples of such innovations include (i) Wecyclers in Nigeria who use low-cost, environmentally friendly cargo bicycles to provide households and businesses with convenient collection services for recyclable waste and (ii) The Rethaka

Foundation's Repurpose Schoolbag initiative which collects and repurposes plastic waste into low-cost, highly visible schoolbags for local disadvantaged students equipped with a small solar panel (Iwuoha JP 2015).

In some African cities, investments have been made in large-scale traditional waste treatment technologies. These include Extrupet in South Africa, which is the first waste PET plastic food-grade bottle-to-bottle recycling plant established in 2015 and has the capacity to recycle over 2.5 million PET bottles per day. In Addis Ababa, Ethiopia, the Reppie 50 MW waste-to-energy (WtE) plant at the Koshe dump marks a significant transition in Africa's management of waste, with the establishment of a large-scale, MSW thermal treatment plant (UNEP 2018).

3.3.1 State of Solid Waste Management in African Cities

The total municipal solid waste (MSW) generated in Africa (in 2012) was estimated to be 125.0 million tonnes a year, of which 81.0 million tonnes were from sub-Saharan Africa (Scarlat et al. 2015). North African countries have a relatively higher per capita waste generation than sub-Saharan countries. The average per capita waste generation in Africa in 2012 was 0.78 kg per day, which is much lower than the global average of 1.24 kg per day. Municipal waste generated was estimated at 125.0 million tonnes a year (Scarlat et al. 2015).

Waste generation in Africa is driven by the following factors in Africa. They include:

I. **Growth in Population.** Population in Africa has been projected to increase from 17% of the global population in 2017 (1.3 billion) to 40% in 2020 (4.5 billion) meaning an increased waste burden on African cities. This will further add more burden to an already strained waste infrastructure (United Nations 2015).
II. **Urbanisation.** Forty per cent of the population in Africa lives in urban areas. This is projected to reach 55.9% of the population by 2050. It is projected that between 2010 and 2025, some African cities will account for up to 85% of the population. This means that the amount of waste generated will increase. However, this will lead to low waste collection rates and open dumping because cities are not keeping up with population growth (United Nations 2015).
III. **Growing Middle Class and Changing Consumption Habits.** There has been established a correlation between waste

management and change in consumption of the growing urban middle class. Currently, African countries' organic waste constitutes more than 65% of total waste compared to 30% for developed countries. (United Nations 2015).

IV. **Economic Development.** The majority of African countries aspire to achieve middle-income county status by 2025 (World Bank 2016). The number of young Africans entering the workforce is estimated at 10–12 million per annum (AfDB 2016). The African Development Bank (AfDB) has initiated a plan called Youth in Africa Strategy 2016–2025, with the aim of creating 25–50 million jobs by driving inclusive growth across the continent and equipping youth to realise their full economic potential. (World Bank 2016). Strategies such as this encourage more people to work in urban areas. The effect is that more waste is generated in the system as more people engage in economic activities. It is even worse where the waste disposal system is inadequate.

V. **Global Trade.** There has been an increase in the illegal exportation of electronic waste (e-waste) to African countries which is dumped in uncontrolled dumpsites causing threats to human health and the environment (UNEP 2005). Available data shows that 125 million tonnes per annum of municipal solid waste (MSW) were generated in Africa in 2012, of which 81 million tonnes (65%) were from sub-Saharan Africa (Scarlat et al. 2015). This is expected to grow to 244 million tonnes per year by 2025. However, with an average waste collection rate of only 55% (68 million tonnes) (Scarlat et al. 2015), nearly half of all MSW generated in Africa, remains within our cities and towns, dumped onto sidewalks, open fields, stormwater drains, and rivers. The average MSW collection rate in sub-Saharan Africa is lower at only 44%, although the coverage varies considerably between cities, from less than 20% to well above 90%. The average MSW collection rate for the continent is expected to increase to only 69% by 2025 (Scarlat et al. 2015).

Good waste collection services are often only found in the city centres, while municipal waste services in suburbs and peri-urban areas are usually poor. The situation is much worse in rural areas where often no formal waste collection services exist. This is due to a lack of or weak legislation and enforcement. Most African countries have waste management legislation but most have not adopted them. There exists limited public awareness of proper waste handling and recycling and general poor household attitudes towards waste management as a service in Africa (UNEP 2018). Current MSW collection services in most African countries are therefore completely inadequate resulting in the leakage of waste into the environment, including the freshwater and marine environment (Godfrey et al. 2019).

An estimated 80–90% of municipal solid waste generated in Africa is recyclable. However, this is disposed of in uncontrolled and controlled dumpsites. Disposal sites are in most cases located in environmentally sensitive areas such as wastelands, wetlands, forest edges, or adjacent water bodies. They often do not have liners, fences, soil covers, and compactors as in most developing countries (Johannessen and Boyer 1999). With so little regard for the opportunity that waste presents as a secondary resource, only 4% of the waste generated in Africa is currently recycled by marginalised informal reclaimers (Godfrey et al. 2019). The private sector is to some extent involved in waste recycling of items such as plastics, paper, and cardboard. This apart from providing a source of livelihood to the waste pickers is helping in removing wastes that could have ended in the environment causing pollution and aesthetic impairment (Okot-Okumu 2012).

Responsibility for waste management is split between the municipality and communities in African urban areas. Other important actors in waste governance include the private sector, civil society, consumers, and the informal sector. In these areas, waste management services are only offered to a few households.

Lack of technical and human resources and restricted funding of public services has led to inadequate urban solid waste management. A low level of awareness of municipal authorities concerning environmental and public health impacts contributes to inadequate urban solid waste management. A good management system relies primarily on good household waste collection, good street cleaning performance, and provision of sanitary landfills (Couth and Trois 2010).

Proper waste management can be achieved if there is an element of public participation. Poor pay, high illiteracy levels, and low GDP per capita influence factors that cause a lack of participation in public management matters.

3.3.2 *Roles/Contributions of the Informal Sector in Waste Management*

Informal waste pickers significantly contribute to waste management by collecting, sorting, trading, and processing waste materials. This is because, in many developing and transitional countries, the infrastructure and organisational systems of waste management are insufficient. The informal waste collectors and recyclers adapt strategies to access waste and further integrate emerging new systems.

The value given to informal sector activities is reflected in local and international names bestowed on the workers involved. The informal sectors consist of individual waste pickers at the lowermost levels, ascending to recycling small and medium-sized enterprises (SMEs), craftsmen and middlemen, brokers, wholesalers, and manufacturing industries. The value of materials usually increases up the chain (Linzner and Lange 2013).

3.3.3 *Social and Economic Opportunities Created by Recycling*

Economic Opportunities

The United Nations has stated that reuse, recycling, and recovery of waste as opposed to dumping in dumpsites and landfills could inject US$8 billion every year into the African economy. By achieving this, secondary resources could be released back into the African economy, growing and strengthening local manufacturing, creating jobs, addressing unemployment, and building local and regional economies. And if done responsibly and sustainably, this can minimise the environmental and human health impacts associated with the current poor solid waste management practices seen across the African continent (UNEP 2018).

Social Opportunities

Recycling creates jobs. An example of job creation in waste management is seen in Burkina Faso where a project for collecting and recycling plastic waste has helped improve the environmental situation and created jobs and generated income for the local community (UNEP 2018). The project is managed by 30 women and supported by around 2000 informal collectors. The recycling centre is also assisted by two technicians. The recycling centre has allowed many people to secure an income, either by collecting plastic waste or by working as full-time employees at the recycling centre (UNEP 2018).

Recycling alleviates poverty. Informal pickers are currently recycling non-harmful wastes. With proper assistance, this could be scaled up and better managed at the neighbourhood level. This can generate income opportunities and improve the livelihoods of men and women in the African continent (UNEP 2018).

3.3.4 *Environment and Social Impact of Solid Waste Mismanagement*

The UNEP Africa Waste Management Outlook, 2018 (UNEP 2018) highlighted certain environmental and social impacts of waste mismanagement. In Africa, open dumping is the most used method of disposal of waste and this has the potential to have serious implications for human health and the environment. Methane and black carbon which are released through the open burning of waste are climate pollutants.

Africa is believed to be a global destination for end-of-life electronics and vehicles which are exported from developed countries in North America, Europe, and Asia. As such, it accumulates e-waste. The informal recycling practices pose potential risks to people and the environment. Lead-acid batteries recycled in informal workshops also lead to lead pollution. Many African countries lack legislative and regulatory measures to manage the final treatment or disposal of healthcare waste. These measures should be put in place to manage the disposal of healthcare waste.

Africa has weak waste collection systems in comparison to the growing plastic waste in the continent. Disposal of plastics is done at uncontrolled and controlled dumpsites which leads to leakage of plastic into the environment posing a significant threat to the environment.

3.4 Waste Management and Carbon Emissions in Africa

Article 3(1) of the Kyoto Protocol provides a limit for the emission of carbon dioxide from greenhouses. Greenhouse gases are emitted by (i) collection and transportation of waste from the combustion of fuel and (ii) dumping of collected waste in unmanaged landfills which leads to pollution of soil and freshwater by leachates and producing methane (CH4), a greenhouse gas 28 times more powerful than CO2. Organic matter accounts for between 50% and 70% of municipal solid waste in low and middle-income countries. Greenhouse gases can be calculated from the decomposition of waste in landfills, incineration, and recycling (Trois and Jagath 2011).

3.4.1 Strategies Used to Promote Carbon Emission Reductions in Africa

There are several strategies that have been implemented in Africa to promote carbon emission reductions, and these include the Clean Development Mechanisms (CDMs) projects, waste recycling, composting, and disposal.

Clean Development Mechanism:
The Clean Development Mechanism has been an instrument in enabling financially unviable projects to become viable based on selling carbon credits. The United Nations has stated that since 2004, CDM has leveraged revenue of about USD315 billion in capital investments for low-carbon projects globally (AfDB, Clean Development Mechanism, 2012). However, most believe that the CDM is not meant for Africa because Africa is not yet prepared in terms of its projects, political commitment, and enabling procedures.

Since 2011, there has been a 90% increase in CDM projects in Africa. There is also the emergence of new market-based mechanisms which have been proposed. This will build on areas where the CDM system has been successful.

In 2021, the AfDB African Carbon Support Programme supported four Bank projects in starting their respective CDM certification processes. The methodology applied was on "Interconnection between Electricity Systems for International Energy Exchange". Its approval by the CDM board makes interconnection projects for clean energy transmission eligible to benefit from the CDM, and thus, from carbon credits. The Ethiopia-Kenya Interconnection Project, used as a case study to support the methodology, has GHG emission reductions estimated at 7 million tCO2/annum (AfDB, Clean Development Mechanism, 2012).

According to the African Development Bank, the projects for CDM validation include the Ethiopia–Kenya Interconnection Project which will see an estimated emission reduction of 7 million tCO2/yr and additional revenue of €350 million over a 10-year crediting period, and the Zina Solar Photovoltaic Power Plant Project which will see an estimated emission reduction of 16,000 tCO2/yr and additional revenue of €800,000 over 10-year crediting period. There is also the Itezhi Tezhi Hydro Power and Transmission Line which will see an estimated emission reduction of 560,000 tCO_2/yr and additional revenue of €28 million over a 10-year crediting period CDM project design and the Lagos Cable Propelled Transit Project which will see an estimated emission reduction of 30,000 tCO_2/yr and additional revenue of €1.6 million over 10-year crediting period.

Recycling:

Recycling has GHG benefits together with energy and water savings. This benefit is highlighted in the Department of Environment and Conservation New South Wales report of 2005 for recycling paper/card, liquid paper board, glass, aluminium, steel, high-density polyethylene (HDPE), and polyethylene terephthalate (PET). The cost of conventional waste disposal is low in comparison to the cost of recycling (Couth and Trois 2010).

Whilst recycling waste does reduce the amount disposed to landfills and effectively saves the municipality costs in the provision and operation of a landfill, this is not reflected in the income from recyclable materials The most important benefit to the municipality will be an extension of the life of existing landfill sites, job creation and cleaner communities (Couth and Trois 2010).

Waste to Energy:

In Africa, the demand for waste to energy facilities, advanced recycling, and bio-waste processing facilities is rising to meet the need for sustainable urban living, improved resource efficiency, and avoidance of further climate change. In Ethiopia, the Koshe landfill has been transformed into a waste-to-energy plant. The plant started operation in August 2018. In waste-to-energy incineration plants, rubbish is burned in a combustion chamber. The resulting heat is used to boil water until it turns to steam, which drives a turbine generator that produces electricity. The Reppie adopts modern back-end flue gas treatment technology to drastically reduce nitrogen oxides, sulphur dioxide, heavy metals, and dioxins produced by the plant. The plant operates safely within the strict emission limits of the European Union. This gives an insight as to how waste-to-energy can positively impact cities in Africa while creating job opportunities and preserving the environment.

Flexible Strategies and Financial Incentives:
Governments in Africa are applying policies in pollution and waste management. This includes economic incentive-based strategies, landfill bans, directive-based legislation, and the introduction of landfill tax. These regulations have to be enforced in order to be of use (Couth and Trois 2010). However, some economists believe that developing countries are not yet ready for financial incentives in relation to greenhouse gas emissions. Some of the reasons given include lack of monitoring and enforcement, lack of constitutional capacity, corruption, and poorly developed markets and legal systems as the major reasons.

3.5 Conclusion and Recommendations

The management of waste in African cities is a challenge that must be tackled head-on. The reason is not far-fetched, the population is increasing. As such, governments must mainstream the relevant policies and regulations to reduce or eliminate the possibility of waste burning, which results in carbon emissions. One way of achieving this is by enacting waste management and decarbonisation legislation and policies nationally and in the municipalities. Strengthening policies and institutions that are harmonised, monitored, and enforced would ensure that the policies go a long way in effectively implementing waste management and decarbonisation.

Improved capacity and awareness among the city's dwellers and active partnerships to solve waste challenges are fundamental. One of the main keys in waste management is having public-private partnerships and participation. This is evident, especially in Dar-es-Salaam and Nairobi whereby different NGOs and CBOs participate in waste collection, recycling, and disposal in the urban council. However, these organisations only focus on the less privileged urban communities which are more than half of the population. Again, countries should set achievable standards and objectives and further involve the locals in problem definition, strategy, and problem solution. They should put in place waste services and technologies that attract investment. Additionally, informal actors in waste management should be incorporated to play a major role in the industry. Governments should include informal actors in order to improve their livelihoods as most are categorised as disadvantaged groups.

Finally, governments should fully embrace recycling technologies and should scale them up by strengthening local and regional end-use markets.

Recycling technologies are already being implemented for wastes such as plastic, paper, glass, metal, oil, e-waste, and organic waste, but could be significantly scaled up through the development and strengthening of local and regional end-use markets.

Abbreviations

AfDB	African Development Bank
CDM	Clean Development Mechanism
CBO	Community Based Organization
EAC	East African Community
E-Waste	Electronic Waste
ECOWAS	Economic Community of West African States
GDP	Gross Domestic Product
GHG	Greenhouse Gas Emissions
MSW	Municipal Solid Waste
NGO	Non-Governmental Organization
SDGs	Sustainable Development Goals
SADC	Southern African Development Community
WtE	Waste to Energy

References

African Development Bank. (2016). Annual Report. Africa Development Bank. https://www.afdb.org/en/documents/document/afdb-annual-report2016-95954

Trois, C., Jagath, R. (2011). Sustained carbon emissions reductions through zero waste strategies for South African municipalities. In Kumar, S. (ed.), Integrated Waste Management - Volume II, 978-953-307-447-4. InTech. http://www.intechopen.com/download/get/type/pdfs/id/18498

Couth R, & Trois C. (2010). Carbon emissions reduction strategies in Africa from improved waste management: A review. *Waste Management*, 30(11), 2336-46. DOI: https://doi.org/10.1016/j.wasman.2010.04.013.

Iwuoha, J.P. (2015). Making Money From Trash – Meet Africa's Top 5 Entrepreneurs in the Waste Recycling Business. https://www.smallstarter.com/get-inspired/africa-top-5-entrepreneurs-in-the-waste-recycling-business/

Johannessen, L. M & Boyer G., (1999). Observations of Solid Waste Landfills in Developing Countries: Africa, Asia and Latin America. The International Bank for Reconstruction and Development. The World Bank, Washington D.C.

Scarlat, N., Motola, V., Dallemand, J. F., Monforti-Ferrario, F., & Mofor, L. (2015). Evaluation of energy potential of municipal solid waste from African urban areas. *Renewable and Sustainable Energy Reviews, 50*, 1269-1286. https://doi.org/10.1016/j.rser.2015.05.067.

Okot-Okumu J. (2012). Solid waste management in African Cities—East Africa. In: Rebellonm LFM, (ed). Waste Management—An Integrated Vision. Rijeka: InTech. http://cdn.intechopen.com/pdfs/40527/InTechSolid_waste_management_in_african_cities_east_africa.pdf

Linzner, R., & Lange, U. (2013, May). Role and size of informal sector in waste management–a review. In *Proceedings of the Institution of Civil Engineers - waste and resource management*, 166(2), pp. 69-83). ICE Publishing.

Godfrey, L., Ahmed, M. T., Gebremedhin, K. G., Katima, J. H., Oelofse, S., Osibanjo, O., ... & Yonli, A. H. (2019). Solid waste management in Africa: Governance failure or development opportunity. *Regional Development in Africa*, 235. DOI: https://doi.org/10.5772/intechopen.86974

United Nations Department of Economic and Social Affairs. (2015). World Population Prospects: Key findings and advance tables. https://population.un.org/wpp/publications/files/key_findings_wpp_2015.pdf

United Nations Environmental Programme. (2005) Annual Report. United Nations Environmental Programme. https://wedocs.unep.org/bitstream/handle/20.500.11822/183/UNEP_Anual_Evaluation_Report_2005.pdf?sequence=1&isAllowed=y%2C%20https

United Nations Environmental Programme. (2018). Africa Waste Management Outlook. United Nations. https://wedocs.unep.org/handle/20.500.11822/25514

World Bank. (2016). Annual Report. World Bank. "World Bank. 2016. The World Bank Annual Report 2016. Washington, DC. World Bank. https://openknowledge.worldbank.org/handle/10986/24985

CHAPTER 4

Innovative Strategies for Decarbonising the Healthcare Sector in Nigerian Cities

Smith I. Azubuike and Adebola Adeyemi

Abstract Africa's health sector emits a sizeable amount of carbon through transportation due to the influx of people into hospitals in the cities seeking quality healthcare, the use of fossil sources to generate power, poor medical waste management, and supply chain reliance on other carbon-emitting sources. The situation is worsened as a result of energy poverty, inefficient transport system, and a weak policy framework. The authors draw data from Cedarcrest Hospitals, Abuja, Nigeria and suggest that goal-setting together with the establishment of innovative approaches will promote decarbonisation of the health sector. We frame the research on ecological modernisation theory by noting that innovative approaches and partnerships, such as the Top-Runner Approach, can reduce carbon emission from healthcare facilities in Nigeria and elsewhere in Africa.

S. I. Azubuike (✉) • A. Adeyemi
Durham Law School, Durham University, Durham, UK

S. I. Azubuike et al. (eds.), *Decarbonisation Pathways for African Cities*, Palgrave Studies in Climate Resilient Societies,
https://doi.org/10.1007/978-3-031-14006-8_4

Keywords Healthcare decarbonisation • Nigerian cities • Ecological modernisation • Climate change • Top runner

4.1 Introduction

Globally, numerous reports promoting decarbonisation highlight the adverse impacts of greenhouse gas (GHG) emissions and promote clean energy transition and initiatives (IRENA 2020; UN 2020). Stabilising GHGs at 1.5 C and reducing the negative effect of climate change requires a critical evaluation of anthropogenic activities and mainstreaming viable GHG emission mitigation techniques (Roeben and Azubuike 2020). One of such areas identified is the health sector's carbon footprint in African cities. On their part, health facilities emit GHG by relying on medical infrastructure that may not be energy-efficient in design and construction, transportation arrangements, energy consumption, and product manufacture, use, and disposal (Loosemore et al. 2011; Karliner et al. 2020).

In a study conducted by EASAC & FEAM (2021, p.2) it was noted that the health sector generates nearly 5% of global carbon emission, with most of this footprint coming from energy use (Healthcare Without Harm and ARUP 2019, p.19; IEA 2021, p.147). Further, Harold (2019) highlights that "If health sector were a country, it would be fifth-largest emitter on the planet". Karliner et al. (2020) find that healthcare facilities are the operational heart of healthcare service delivery and are used to protect, care for, and save lives. Nevertheless, the facilities relied on constitute a significant source of carbon emissions. These carbon releases contribute to global warming and in turn, cause health challenges such as respiratory diseases and other climate change-induced illnesses (Watts et al. 2020).

Although there is no exhaustive chronicle of the exact amount of GHG emission from healthcare facilities in Africa, scholars note that Africa's health facilities contribute to carbon emission (Adair-Rohani et al. 2013; Salas et al. 2020). The reason is that rural healthcare facilities in the region are in a poor state (IEA 2014), such that rural dwellers flock to cities to seek better medical care. This trend puts additional pressure on the use of medical facilities in the metropolises and translates to increased transport use in the cities. Transport and operational medical vehicles and supply chain logistics use fossil fuels that release GHGs that contribute to global warming.

According to the World Health Organisation (2015), one out of four healthcare facilities in Sub-Saharan Africa (SSA) is without electricity access. Blimpo and Cosgrove-Davies (2019, p.13) suggest that nearly 60% of all health centres in SSA do not have access to electricity, with just 34% of hospitals and 28% of health clinics having reliable access. Although limited access to electricity highlights the energy poverty situation in SSA, the challenges exceed the access shortfall. The poor state of the electricity sector in some African countries arises from persistent reliability challenges, inadequate consumption, prices that are not cost-reflective, and utilities in financial distress (Blimpo and Cosgrove-Davies 2019, p.12). As a result, these health facilities try to meet their electricity needs by relying on fossil fuel sources to generate power. Besides, some of the waste generated is not adequately disposed of, and certain products and chemicals such as anaesthetic gases used in delivering health services in cities emit GHGs, resulting in pollution to the environment. The climate impact of these gases ranges between 130 kgCO2e/kg (sevoflurane) and 2540 kgCO2e/kg (desflurane), and most of them are absorbed into the atmosphere (Fletcher 2019).

We frame the research on ecological modernisation (EM) theory to discuss the relevant literature and highlight the importance of reducing GHG emissions from the healthcare sector. The framework supports transforming a carbon-based system into a decarbonised system (Mol 1997; Buttel 2000; Mol and Spaargaren 2000; Murphy and Gouldson 2000; EASAC and FEAM 2021). We utilise a literature review and qualitative data from Cedarcrest Hospitals, Abuja, Nigeria, to identify the energy efficiency approach of hospitals in Nigeria. Our analysis reveals that as a result of epileptic power supply, the hospital emits substantial carbon due to its reliance on diesel generators as an alternative power source and a waste management system that needs upscaling. Against this background, this study advances approaches that promote decarbonisation of healthcare services and facilities in Nigeria, primarily as rural and urban dwellers rely mostly on healthcare facilities in cities for quality healthcare. We consider the impact of carbon emission on the environment and human health to advance our low-carbon health sector's objective. We note that innovative approaches and partnerships, such as the Top-Runner Approach, can reduce carbon emissions from healthcare facilities and be mainstreamed into the healthcare sector. Climate-smart strategies will promote the reduction in the carbon footprint of healthcare facilities and stimulate climate-friendly health products while preparing the healthcare sector for future climate challenges.

4.2 Conceptualising Healthcare Decarbonisation Through Ecological Modernisation

4.2.1 *Ecological Modernisation: Concept and Perspective*

The idea of ecological modernisation (EM) emanated from broad social theory and transformed into environmental-social science and helped shape the idea around environmental practices (Mol and Spaargaren 2000). It was first discussed by Jänicke (1985) and adopted by Huber (2000) to examine the extent to which EM can promote sustainable development. EM involves measures taken to promote environmental innovation and its distribution and how these actions complement eco-efficient innovation. Specifically, the core of EM lies in prevention, innovation, and structural transformation to reduce human impact on the environment while still pursuing economic growth (Adua et al. 2021). Thus, it emphasises that environmental sustainability can co-exist with economic growth by using innovative and technological approaches (Mete and Xue 2021; Mojumder and Singh 2021).

The driving forces behind EM, a technology-based and innovative method to ecological policy, are the challenge posed by climate change, the market rationale of modernisation, and the quest for innovation combined with the market prospect of global environmental needs (Jänicke 2008). In addition, smart government regulations and the business risk for polluters within the context of multi-layered environmental control also account for the push for EM.

In the context of a low-carbon framework, Howes (2009) notes that EM provides a foundation that encourages transformation from a carbon-based to a decarbonised system and suggests guides to better policy design to transition to a sustainable world. This transition could come through cleaner technology, efficient use of energy, products, or transport to reduce carbon emissions (Jänicke 2008). It could also be through the formulation of policy frameworks designed to address a particular institution's challenges in a bid to achieve a balance between man, the ecosystem, and the economy (Gibbs 2000).

EM utilises a set of social postulations with analytical and normative aspects related to society's change and policy to address environmental issues (Mol 1997; Murphy and Gouldson 2000). An essential notion of EM relates to environmental readaptation for ecological protection, and reducing healthcare's footprint fits this description. This is because the healthcare sector requires increases in resource and energy efficiency and product and process inventions like sustainable supply chain management to reduce carbon emissions. As such, low-carbon practices in delivering health services in African cities are in sync with the ecological modernisation framework and help drive climate change agendas for a sustainable future and healthy living.

Mol and Sonnenfeld (2000) identify that EM focuses on current and planned ecological changes in institutional designs, social customs, and policy discourses to protect the basis of societies' sustenance. Although the diverse analytical aspect of EM theory exists (Buttel 2000; Mol and Spaargaren 2000; Murphy 2000), the normative aspect appears more consistent and is often used as a framework to design and describe a practical political programme to readdress environmental policymaking (Mol 1997; Murphy 2000).

EM recognises the role of science and technology in mitigating ecological problems and identifies the essence of market dynamics and economic agents as conveyors of environmental restructuring. It identifies the need for state intervention in the market to attain economic growth and environmental protection (Malmborg and Strachan 2005). Doing this entails establishing high environmental rules to convey essential preferences for invention and moving beyond command-and-control instruments in green policy to using different pioneering policy instruments such as economic instruments, voluntary agreements, and market solutions. This practice could 'ecologise' the economy and 'economise' the environment. This is a win-win situation for the society, improving environmental performance, and economic competitiveness at the macro and micro levels (Malmborg and Strachan 2005) and decarbonising healthcare delivery.

4.2.2 Reviewing the Literature on EM

Scholars have highlighted the need to adopt modern approaches in dealing with environmental issues in society. But before now, sceptics argue that EM will deindustrialise the economy (Mol 1995; Spaargaren and Mol 1992). This perception may have driven the United States' and other countries' perspectives on carbon reduction. Analysis about EM suggests that it is a social theory in which growth arises from its connection to broad political and economic factors, much of which exists outside sociology and environmental sociology (Buttal 2000).

EM has been used in the agricultural sector to examine whether agro-ecological methods can contribute to the future demand for food production, particularly in emerging economies. The research found that strategies for agro-ecology can substantially contribute towards 'feeding the world', thus contributing to a genuinely green revolution (Horlings and Marsden 2011). Within the context of sustainable building, EM has been used to account for how zero-carbon home agendas have evolved in England and how social, political, and economic forces interact in providing environmental services and products (Lemprière 2016). Similarly, Jensen and Gram-Hanssen (2008) note that EM has influenced and has now been integrated into the construction sector's sustainable building concept.

In developing renewable energy, Toke (2011) examines EM and renewables and notes that economic development and ecological protection through renewables can be combined where political will and public support exist, not forgetting to undertake a precise identity analysis. On their part, Malmborg and Strachan (2005) have used EM to conceptually drive discussions in carbon emission trading and advance decarbonisation's economic and environmental essence. Relatedly, Howes (2009) utilises EM to reconstruct the challenge facing carbon trading in Australia and identify that it arises from inefficiency, requiring institutional and technical design as an adequate response. Huber (2000) examines EM from how it can decarbonise industrial activities to achieve a climate plan and ensure a sustainable future. In Africa, Rufus (2014) applied EM as a framework to analyse the environmental damage caused by oil pollution in the southern part of Nigeria and shows how the concept informs the need for the oil industry to reduce carbon emissions in the area.

An essential aspect of EM is introducing low-carbon criteria into the production, consumption, and delivery process of services such as healthcare. This emphasis assigns a crucial role in the health sector to protect the environment in healthcare delivery. We note that existing literature on healthcare and low-carbon energy is fragmented, with scholars focusing on how climate change impacts health (Watts et al. 2020). They also consider how healthcare delivery contributes to climate change (Karliner et al. 2020; Pichler et al. 2019) and what developed climes are doing to reduce healthcare emissions in their jurisdictions (Salas et al. 2020; HCWH 2018).

Various scholars have approached the concept of EM from several low-carbon perspectives. We contribute to the discourses from the unique dimension of healthcare provisioning. No known research points to Africa's health sector's carbon footprint, but the health sector emits carbon through its supply chain and service delivery. Thus, policy intervention will be needed to reduce the carbon footprint from the sector, especially in cities. Since EM encompasses the use of strategic environmental management policy innovation to minimise pollution and waste (Buttel 2003), it provides a theoretical foundation to advance our argument for the decarbonisation of the health sector in Africa for climate action and wellbeing.

4.3 Carbon Emission in Africa's Healthcare Facilities

4.3.1 Analysis of Carbon Emission Data from a Private Health Facility in Nigeria

Decarbonising healthcare starts with recognising and appraising the carbon footprint from health facilities (Salas et al. 2020). In this regard, we obtained the following data from Cedarcrest Hospitals, Abuja, Nigeria ("the Hospital") through an interview with an official at the Hospital. The data and responses suggest that the health facility consumes approximately 90,000 kWh of energy per month and does not generate renewable energy. The Hospital relies on the Government-approved waste collection agency and this waste management approach is in line with the existing procedure for waste collection in the Hospitals' location. This waste management system requires upscaling taking into account the support that could be given to the waste collectors to improve their own waste management processes as the collectors' recycling capacity and capability is not clear.

With respect to sourcing energy-efficient products from third parties, the response obtained from the Hospital suggests that the Hospital sources and uses energy-efficient products. Although there is no written policy on this, the procurement staff are aware of the importance and cost benefits of purchasing energy-efficient products. Regarding energy generation, our analysis of the energy efficiency approach of the Hospital shows that it emits more carbon due to its complete reliance on diesel generators as an alternative to the epileptic power supply to power the Hospital and their waste management system.

Globally, there is a lack of a standardised approach to calculating and tracking climate footprint in the health sector (HCWH & ARUP 2019, p.39). The Health Ministry in Nigeria has little capacity to calculate and track the climate footprint of the healthcare sector. As such, we scored the Hospital's performance based on how their activities enhance the decarbonisation agenda under the various themes shown in Table 4.1. The score was allocated using the following scoring system: 0–2 (Not appropriate); 3–4 (Somewhat appropriate); 5–6 (Slightly appropriate); 7–8 (Appropriate); 9–10 (Very appropriate). We identified the waste management system of the Hospital as Somewhat Appropriate because despite its awareness of relevant laws requiring proper disposal of medical waste, it still does not have a clear decarbonisation strategy for such waste. It rather depends on the municipal waste authority who often burns collected waste (Ferronato and Torretta 2019). Regarding the sourcing of medical equipment, the score of 'Appropriate' is dominant due to the Hospital's use of environmentally friendly products. The energy generation and consumption were scored 'Not Appropriate' because the energy consumption of the Hospital is high while the source of generation is mainly from fossil fuel sources.

4.3.2 The Impact of Carbon Emission on Health and the Environment

The impact of decarbonisation and moving to clean energy is important for the healthcare sector, as it will contribute to reducing the sector's carbon footprint. It will also promote sustainable development, improved economic activities, better health outcomes through a reduction in the burden of diseases, Cronk and Bartram (2018) and a cleaner environment. It has been identified that Africa and particularly Nigeria may be susceptible to most of the adverse effects of climate change (World Bank 2010). Ironically, Nigeria is a stark reflection of the ill-preparedness of the African continent (Terr-Africa 2009). With respect to day-to-day health, the global negative effect of carbon emissions is far-reaching (NPI 2006) with reports suggesting that air pollution is linked to around 2.5 million deaths globally and 500,000 in SSA (IEA 2019, p.102; Healthcare Without Harm and ARUP 2019, p.36). The most susceptible to these effects are children with undue prolonged exposure to harmful substances resulting

Table 4.1 Energy use from Cedarcrest Hospitals, Abuja, Nigeria

S/ No	*Theme*	*Question*	*Cedarcrest's response*	*Score 0–10*
1	Waste management	What is the process of waste management and procedure for disposal?	Mostly contracted out through the municipal authority (AEPB) in charge of waste collection and disposal.	4/10
		Is the Hospital aware of any law mandating the proper disposal of medical waste?	Yes, it is.	
		Does the Hospital have a policy on the proper disposal of medical waste?	Will need a revision of the entire process for waste management and disposal.	
2	Sourcing of medical products/ materials	Does the Hospital consider environmentally friendly products when sourcing for materials/ input?	Yes it does	7/10
		Are energy saving bulbs purchased?	Yes	9/10
		Are the computers energy-efficient?	A large percentage are energy-efficient	6.5/10
		Are the lab equipment energy-efficient?	A major percentage are energy-efficient	7.5/10
		Does the Hospital have a policy that promotes the purchase and use of energy-efficient products?	Awareness is above average, but no written document about it	6.5/10
3	Energy generation	Does the Hospital have a clean energy generation system installed? If yes, Is it solar, hydro, or gas?	No. Diesel Power Generators are used as alternative energy generation system	0/10
4	Energy consumption	What is the total electricity consumed by the Hospital per month (kWh)?	Approximately 90,000 kWh per month	1/10
		Generating set capacity	Maximum 500 KW	
		What is the fuel consumption rate?	18,000 litres of diesel is consumed by the Diesel Power Generators per month (@ 220 naira (N) per litre = N3, 960,000 per month)	

in negative health outcomes including impairment of lung development and cognitive ability, immune system alteration, and increased risk of vitamin D deficiency (Odusanya et al. 2014, p.59; UNFCCC 2018).

Carbon emissions and by-products contribute to climate change and can complicate asthma attacks, respiratory illnesses, and damage the lungs. In substantially high concentrations, certain carbon emissions can result in serious lung damage resulting in chest pain, lung cancer, and eventual death. Methane, for instance, is an asphyxiant notable for its ability to displace oxygen—at 18% displacement, it is capable of inducing asphyxia. In more serious cases, respiratory ailments could force miscarriages. For context, certain areas of Nigeria have an overwhelming prevalence of respiratory illnesses as common denominators to certain demographics. Figure 4.1 highlights how carbon emissions trigger climate change, which then impacts human health.

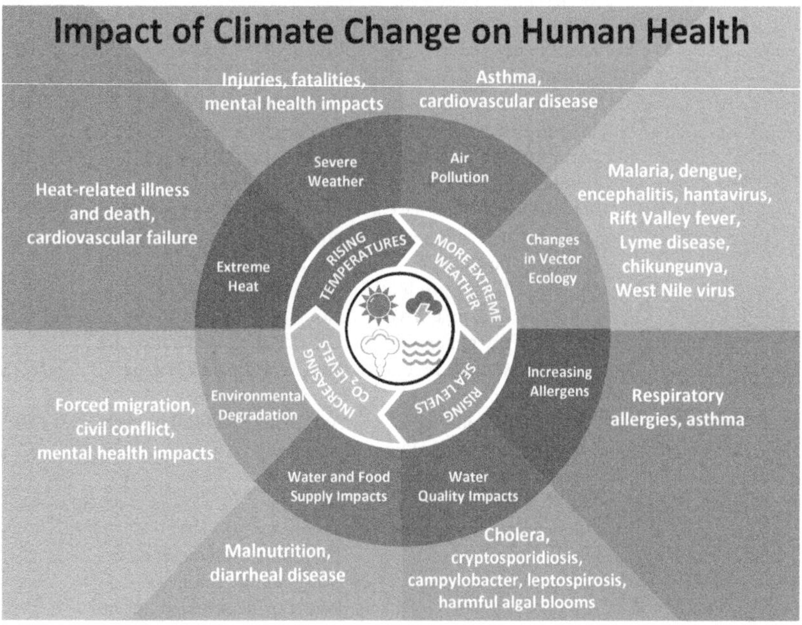

Fig. 4.1 The impact of climate change on human health (Source: Health Care Without Harm 2019)

4.4 Decarbonising the Healthcare Sector

4.4.1 *Approaches*

At the core of EM is how innovative approaches could help countries to decarbonise their economies. We discuss some eco-efficient strategies to cut carbon emissions from products and supplies, transport, renewable energy systems, and low-carbon waste management practices in the healthcare sector.

4.4.1.1 The Top-Runner Approach

This eco-efficient approach emanated in Japan in the 1990s to assess energy use and enhance the effectiveness of end-products that consume energy. As part of Japanese Energy Conservation Law, the programme set varying compulsory standards of efficiency for different equipment, appliances, and automobiles based on the most efficient products—"Top Runner" products—on the market (Kimura 2012). The most efficient energy product in the market sets the Top Runner standard, albeit consideration is given to products with technological potential and capacity for future efficiency improvements. This promotes the identification, development, and adoption of energy-efficient products.

The programme, which started with a small number of products, has been increased and is now deemed one of the main pillars of Japanese climate policy (Jänicke 2008). The method of assessing compliance with the standard is through corporate average product sales, which relies on a dedicated institution that determines standards and an advisory committee drawn from various aspects of society (Kimura 2012).

As already highlighted, the health sector uses ventilators, air conditioners, fridges, electric bulbs, computers, anaesthetic gases, and medical supplies that emit GHG. Most of this equipment is imported into Africa. Allowing only the most energy-efficient equipment that qualifies as 'Top Runner' products, for use in hospitals is in tandem with the EM. This approach promotes the use of the most energy-efficient products to cut emissions in cities where the most sophisticated healthcare facilities are located. However, this will require smart environmental legislation, policies that promote low-carbon procurement for pharmaceuticals and medical gadgets, recognition of environmental responsibility by the private sector, institutions that determine the most energy-efficient products, and the political will to implement it.

Unlike command and control, this approach advances a prescriptive and performance-based strategy that can stimulate the manufacturing of energy-efficient products with low-carbon potentials for use in hospitals. This avoids 'environmental dumping' of old medical equipment and supplies in Africa (Andersen et al. 2018). There is, however, a cost implication on healthcare consumers as energy-efficient products generally have a larger upfront cost compared to less energy-efficient products and this may result in higher cost of health services. A way to address this is for the Government to create an enabling environment to promote the local development of energy-efficient products and improve the technology transfer framework. Also, the Government can assist by providing direct and indirect subsidies such as tax holidays, reduced customs duties, and access to cheap credit for such development efforts. The effectiveness of these interventions will need to be carefully planned, bearing in mind that the inputs required for most energy-efficient devices are natural mineral resources sourced mainly from resource-rich developing countries (Sembiring 2020, p.9).

The effectiveness of the recommendations presented in this section can be further promoted in line with the obligations of countries captured in international treaties and conventions, which set out climate-based responsibilities and actions. To demonstrate Nigeria's attitude and appetite for reducing the impact of climate change, Nigeria has signed and ratified supranational instruments on climate change including the United Nations Framework Agreement for Climate Change (UNFCC), the Kyoto Protocol, the Global Gas Flare Reduction Partnership, and the Zero Routine Flaring Initiative (Olashore 2019). While at the domestic level, several regulations and policies are in place concerning climate change, including the National Policy on Climate Change (2013), the National Gas Policy (2017), the Flare Gas Regulations (2018), and the National Renewable Energy and Energy Efficiency Policy (2015) (Grantham Research Institute 2021; NESREA 2021).

4.4.1.2 Other Measures

Other measures include:

a. *Renewable energy*

The utilisation of renewable energy sources has been identified as a means of minimising GHG emissions (EASAC & FEAM, 2020). Africa offers significant potential for renewable energy deployment, given its enormous sunshine, wind, and hydropower (IRENA 2015). As noted earlier, one out of four healthcare facilities in SSA has no electricity access. Only 28% of health facilities and 34% of hospitals had reliable access to electricity (Blimpo and Cosgrove-Davies 2019, p.12). As such, these health facilities meet their electricity needs by using fossil fuel sources to generate power (UN Foundation and SEforALL 2019, p.10). The result is that health facilities, as seen in Table 4.1, increasingly rely on carbon-emitting sources to meet their energy needs.

Reversing this trend requires the massive deployment of solar PV systems, wind turbines, and reliance on hydropower—for hospitals close to water sources—to generate electricity. In addition, the government could facilitate the process by providing incentives for public and private hospitals that install renewable energy systems. For instance, hospitals can earn money back from their energy supplier under the government Feed-in-Tariff scheme when they generate energy from renewable energy sources and export the surplus to the grid (7 Energy 2021). The government could also create partnerships between private investors and hospitals and remove internal barriers to renewable energy investment. This will allow hospitals to finance renewable energy projects through direct ownership—secure lending, vendor financing, and leasing—with a renewable energy asset on site or via third-party ownership, using a power-purchase agreement between the hospital and the asset owner.

b. *Greening the transport system*

The poor state of rural health facilities incentivises the migration to cities in search of quality health service or medical prescriptions. Travel to towns both for hospital staff and patients is by public or private transportation, which increases the consumption of fossil fuel. This increases the amount of carbon emitted through transportation in the city. One way of curbing this emission is to introduce incentives for the staff of these health facilities to share rides and use public transport. Further, residence near the healthcare facility ought to be made available at low rates to healthcare workers. Hospital management can ensure onsite accommodation is offered as part of the employment package for staff. The Seattle Children Hospital example is instructive here. The hospital has obligations with the government authorities and its neighbours to lower the number of single-occupancy-vehicle (SOV) travel by its staff. It also uses pay-to-park to discourage people from using their cars, subsidise all other travelling types, and offer amenities and services to help employees make smart commute choices. This strategy reduced the SOV travel rate from 73% in 1995 to 38% in 2015 (Practice Greenhealth 2020).

Pay-to-park options can also apply to patients attending hospital appointments, while telemedicine can be used for preventive care and treatment. Telemedicine involves delivering healthcare services from a distance through electronic means for "the diagnosis of treatment, prevention of disease and injuries, research and evaluation, and education of health care providers" (Kirsh et al. 2015) to improve health. This strategy can help to reduce the influx of people to the city for medical care. The challenge with this strategy would be low internet penetration and energy poverty in rural areas, which can be addressed through renewables in off-grid areas. There is also the challenge of the limited nature of diagnosis that may be carried out when the patient is not physically present. Additionally, the government can partner with investors to provide electric scooters and bikes for hospital staff and the patients. This equipment can be recharged using electric charging stations provided at the hospitals, powered by renewable energy sources.

Low-carbon transport strategies in the design of transport projects such as reducing the need to commute, encouraging cheaper transport means, and enhancing system management by reducing congestion in cities' transport capacity could reduce carbon emissions from cars (Agarana et al. 2017). This approach requires the government to transform the energy system to be less dependent on fossil fuel and coal, put a carbon limit on cars, and deploy green transit buses. The private sector could also help by implementing some of the strategies highlighted above.

c. *Supply chain and waste management approaches*

Emission from the procurement of goods and services constitutes a substantial part of a hospital's carbon footprint. But sustainable procurement approaches, policies and practices can be a critical force for decarbonising the supply chain and attaining climate-smart healthcare. For instance, England's Carbon Reduction Strategy of 2009 provides professionals with tools, methods, and guidance to identify and understand hospitals' carbon reduction opportunities (HCWH 2018). Hospitals in Africa could adopt this strategy by using disposable paper boxes, instead of using plastic bags and boxes, as part of their green procurement policy. They can also use reprocessed surgical devices, use alternatives to one-use products, reduce emissions from anaesthetic gases by replacing Desflurane with Sevoflurane if possible, and decrease or eliminate the use of nitrous oxide (HCWH 2018). In managing medical waste, hospitals could use biodigestion instead of burning and food waste to be composted and used as compost for gardening/farming (Stringer 2020). Our evaluation of the waste management practice at Cedarcrest Hospitals indicates that their current strategy could be improved if the Hospital internalises its medical waste disposal arrangement. Reliance on a public institution to dispose its medical waste may not be the best, as public institutions in some parts of Africa are known to dispose waste improperly by burning it, which impacts the climate. However, where an obligation exists to use public institutions to dispose of medical waste, the institution could consider biodigestion as an advanced method of waste management.

4.4.2 *Partnership*

Low-carbon systems require technical expertise, financing, institutional capacity, and governance structures to function, most of which are lacking in Africa but can be accelerated through partnerships with regional and multinational organisations. Collaborations with organisations such as Power Africa, Sustainable Energy for All, etc., is critical in facilitating a net-zero pathway for Africa's health sector through renewables. Maharashtra in India installed 407 hybrid solar PV systems for health facilities between 2008 and 2015 through a partnership with Sustainable Energy for ALL (Salas et al. 2020).

African health facilities could also partner with pharmaceutical companies to reduce medical waste's environmental impacts and obtain green healthcare projects financing. An instance is the Pfizer Sustainability Bond Framework through which Pfizer issues Green, Social, and Sustainability Bonds to finance or refinance projects with environmental advantages (Pfizer 2020). Some multinational organisations could also partner with African health facilities to formulate blueprints for low-cost, health-promoting systems that reduce disease burden, mitigate GHG emissions and local pollution, and adapt climate-smart healthcare practice systems (World Bank Group 2017). Hospitals can also obtain funding through The Green Climate Fund etc to support their decarbonisation process. Furthermore, policy-makers in Africa need to work in partnership with healthcare facilities in developing financing and mitigation measures which must involve actions that facilitate the realisation of the sector's decarbonisation objective as part of an integrated action plan for all sectors of the economy.

4.4.3 Planning and Overcoming Barriers

System change at scale is daunting; thus, the planning process should involve assessing the direct and indirect pollution sources in hospitals. The state should intervene by establishing environmental strategies and policies and adding renewable energy into the energy mix. Integrating decarbonisation approaches into planning processes, incorporating climate analysis and interventions into hospital safety assessments and planning, switching off appliances, and investing in energy-efficient equipment and devices will be beneficial. Barriers such as weak institutional capacity and expertise, inability to finance projects, social barriers, and absence of governance mechanism/strategies exist within the African setting and should be addressed.

Overcoming these barriers would require institutional strengthening, policy formulation, and the political will to address the carbon footprint from the healthcare sector. It will also include broad transformative steps such as cutting demand through preventive care, electrifying hospitals with renewables, opting for medical products and equipment with lower carbon footprints, and using telemedicine to lower the need to travel (Salas et al. 2020). Further, targeted enlightenment efforts to address cultural issues that carry adverse impact on the environment need to be pursued with vigour. For instance, smoke from burning wood is seen as beneficial for reducing bad odours, for improving the taste of food, and for repelling small reptiles and insects (IEA 2019, p.104). However, these actions contribute to degrading the environment and carry health consequences for individuals. If individuals are enlightened on the dangers of burning wood and if innovative solutions can be introduced, this will reduce the pressure on the healthcare sector to treat pollution-related illnesses and in turn, this will reduce the carbon emission from the healthcare sector. Also, there is a need to partner with donor agencies and take advantage of programmes that support carbon reduction in the health sector.

Notwithstanding, it is essential to highlight that some of the recommendations advanced in this paper may be difficult to implement immediately or to the extent proposed as a result of the existing legal, political, and economic challenges that exist in Nigeria. A wide-scale effort and review is necessary to surmount these challenges so that the proposed decarbonisation approaches can take root and flourish.

4.5 Conclusion

This study examined the health sector's carbon footprint, with a particular focus on Nigeria. Despite the focus on Nigeria, other African countries can apply some of the recommendations in their entirety or with slight modifications. This possible application arises from similarities in socio-economic and political factors in Africa, such as energy poverty and limited energy access in healthcare facilities, dependence on fossil fuel for electricity generation, and the lack of infrastructure and political will to facilitate a decarbonised healthcare sector. We obtained data on energy use and carbon emissions from Cedarcrest Hospitals to highlight the energy efficiency approaches of health facilities in Nigeria. We find that the healthcare sector, in general, emits a significant amount of carbon as it depends on fossil fuel sources, anaesthetic gases, medical products, and equipment, and through its carbon-intensive supply chain and waste management system. However, we note that Africa's healthcare setting has a minor per capita carbon footprint on the environment compared with the West. But the general intensity can be significant, and the capacity to minimise energy usage, modify procurement practices, or introduce far-reaching changes is weak in Nigeria.

We frame the study on ecological modernisation theory, which emphasises that environmental sustainability can co-exist with economic growth by using innovative and technological approaches (Mete and Xue 2021; Mojumder and Singh 2021). We build on this to advance the importance of reducing GHG emissions from the healthcare sector. Health facilities in Nigeria lack reliable electric power supply and an appropriate framework for decarbonising the sector. Strategies such as the Top Runner approach, a renewable energy strategy, green transport system, supply chain and waste management, and partnership to overcome barriers to decarbonisation represent possible approaches. These climate-smart strategies will assist in reducing the carbon footprint of hospitals in Nigeria and stimulate the development of climate-friendly health products while preparing the sector for any future climate challenge and supporting economic growth. This approach can be rolled out for other African cities with similar socio-economic constraints, thereby promoting the African cities' decarbonisation agenda.

References

7 Energy. (2021). Funding and financial incentives available for renewable energy systems. Available from http://www.7energy.co.uk/grants.html (Last accessed 15 July 2021)

Adair-Rohani, H., Zukor, K., Bonjour, S., Wilburn, S., Kuesel, A. C., Hebert, R., & Fletcher, E. R. (2013). Limited electricity access in health facilities of sub-Saharan Africa: A systematic review of data on electricity access, sources, and reliability. *Global Health: Science and Practice*, *1*(2), 249-261.

Agarana, M. C., Bishop, S. A., & Agboola, O. O. (2017). Minimising carbon emissions from transportation projects in Sub-saharan Africa cities using mathematical model: A Focus on Lagos, Nigeria. *Procedia Manufacturing*, *7*, 596-601.

Andersen, S. O., Ferris, R., Picolotti, R., Zaelke, D., Carvalho, S., & Gonzalez, M. (2018). Defining the Legal and Policy Framework to Stop the Dumping of Environmentally Harmful Products. *Duke Environmental Law & Policy Forum*, 29(1).

Adua, L., Clark, B., & York, R. (2021). The ineffectiveness of efficiency: The paradoxical effects of state policy on energy consumption in the United States. *Energy Research & Social Science*, 71, 101806.

Blimpo, Moussa P. & Cosgrove-Davies, Malcolm (2019). Electricity Access in Sub-Saharan Africa: Uptake, Reliability, and Complementary Factors for Economic Impact. Africa Development Forum; Washington, DC: World Bank.

Buttel, F. H. (2000). Ecological modernisation as social theory. *Geoforum*, 31(1), 57-65.

Buttel, F. H. (2003). Environmental sociology and the explanation of environmental reform. *Organization & Environment*, *16*(3), 306-344.

Cronk, R., & Bartram, J. (2018). Environmental conditions in health care facilities in low-and middle-income countries: coverage and inequalities. International Journal of Hygiene and Environmental Health, 221(3), 409-422.

European Academies' Science Advisory Council and Federation of European Academies of Medicine (2021). Decarbonisation of the Health Sector: A Commentary by EASAC and FEAM, Available from https://easac.eu/fileadmin/PDF_s/reports_statements/Health_Decarb/EASAC_Decarbonisation_of_Health_Sector_Web_9_July_2021.pdf.pdf (Last accessed 08 October 2021)

Ferronato, N., & Torretta, V. (2019). Waste mismanagement in developing countries: A review of global issues. *International Journal of Environmental Research and Public Health*, 16(6), 1060.

Fletcher, E. A. (2019, 10 September). Health Care Climate Footprint Is 4.4% Of Global Emissions, Larger than Japan Or Brazil. Available from https://healthpolicy-watch.news/health-care-climate-footprint-is-4-4-of-global-emissions-larger-than-japan-or-brazil/ (Last accessed 22 December 2020).

Gibbs, D., (2000). Ecological modernisation, regional economic development, and regional development agencies. *Geoforum*, 31(1), 9-19.

Grantham Research Institute on Climate Change and the Environment (2021). Climate Change Laws of the World. Available from https://climate-laws.org/legislation_and_policies?from_geography_page=Nigeria&geography%5B%5D=130&type%5B%5D=executive (Last accessed 17 July 2021)

Health Care Without Harm (HCWH). (2018). Reducing the carbon footprint of healthcare through sustainable procurement. Available from https://noharm-europe.org/articles/news/europe/reducing-carbon-footprint-healthcare-through-sustainable-procurement (Last accessed 22 December 2021)

Healthcare Without Harm & ARUP (2019). Health Care's Climate Footprint How the Health Sector Contributes to the Global Climate Crisis and Opportunities for Action, Climate-smart health care series Green Paper Number One, Available from, https://noharm-global.org/sites/default/files/documents-files/5961/HealthCaresClimateFootprint_092319.pdf (Last accessed 10 October 2021)

Horlings, L. G., & Marsden, T. K. (2011). Towards the real green revolution? Exploring the conceptual dimensions of a new ecological modernisation of agriculture that could 'feed the world'. *Global Environmental Change*, *21*(2), 441-452.

Howes, M. (2009, September). A climate of change: Ecological modernisation and the politics of carbon trading in Australia. In Australasian Political Studies Association Conference, Macquarie University, Sydney (pp. 27-30).

Huber, J. (2000). Towards industrial ecology: sustainable development as a concept of ecological modernisation. *Journal of Environmental Policy and Planning*, 2(4), 269-285.

International Energy Agency (IEA) (2014). PV Systems for Rural Health Facilities in Developing Areas: A completion of lessons learned. Available from https://iea-pvps.org/wp-content/uploads/2020/01/IEA-PVPS_T9-15_2014_PV_for_rural_health_facilities.pdf (Last accessed 20 September 2021)

IEA (2019) Africa Energy Outlook, Available from https://www.iea.org/reports/africa-energy-outlook-2019 (Last accessed 02 September 2021)

IEA (2021) Tracking SDG7: The Energy Progress Report 2021, Available from https://iea.blob.core.windows.net/assets/b731428f-244d-450c-8734-af19689d7ab8/2021_tracking_SDG7.pdf (Last Accessed 08 October 2021)

International Renewable Energy Agency (IRENA) (2015). Africa 2030: Roadmap for a Renewable Energy Future. Available from https://www.irena.org/-/media/Files/IRENA/Agency/Publication/2015/IRENA_Africa_2030_REmap_2015_low-res.pdf (Last accessed 14 December 2021)

IRENA (2020). Global Renewables Outlook: Energy Transformation 2050. International Renewable Energy Agency, Abu Dhabi.

Jänicke, M. (1985). *Preventive environmental policy as ecological modernisation and structural policy.* WZB Berlin: IIUG.
Jänicke, M. (2008). Ecological modernisation: New perspectives. *Journal of Cleaner Production*, 16(5), 557-565.
Karliner, J., Slotterback, S., Boyd, R., Ashby, B., Steele, K., & Wang, J. (2020). Health care's climate footprint: The health sector contribution and opportunities for action. *European Journal of Public Health, 30*(5), ckaa165-843.
Kimura, O. (2012). The Role of Standards: The Japanese Top Runner Program for End-Use Efficiency. Historical Case Studies of Energy Technology Innovation. in Grubler A., Aguayo, F., Gallagher, K.S., Hekkert, M., Jiang, K., Mytelka, L., Neij, L., Nemet, G. & C. Wilson. *The Global Energy Assessment* (Cambridge: Cambridge University Press).
Kirsh, S., Su, G. L., Sales, A., & Jain, R. (2015). Access to outpatient specialty care: solutions from an integrated health care system. *American Journal of Medical Quality, 30*(1), 88-90.
Lemprière, M. (2016). Using ecological modernisation theory to account for the evolution of the zero-carbon homes agenda in England. *Environmental Politics*, 25(4), 690-708.
Loosemore, M., Carthey, J., Chandra, V., and Mirti A. (2011). Climate change risks and opportunities in hospital adaptation. *International Journal of Disaster Resilience in the Built Environment*, 2(3), 210-221.
Mete, S., & Xue, J. (2021). Integrating environmental sustainability and social justice in housing development: Two contrasting scenarios. *Progress in Planning*, 151, 100504.
Mojumder, A., & Singh, A. (2021). An exploratory study of the adaptation of green supply chain management in construction industry: The case of Indian Construction Companies. *Journal of Cleaner Production*, 295, 126400.
Mol, A. (1995). The Refinement of Production: Ecological Modernization Theory and the Chemical Industry. The Hague: cip-data Koninklijke Bibliotheek).
Mol, A. P. J. (1997). Ecological modernisation: Industrial transformations and environmental reform. In Redclift, M., Woodgate, G. (eds.), *The international handbook of environmental sociology* (pp. 138-149). London, England: Edward Elgar.
Mol, A. P., & Sonnenfeld, D. A. (2000). Ecological modernisation around the world: An introduction. *Environmental Politics*, 9 (1), 1-14.
Mol, A. P., & Spaargaren, G. (2000). Ecological modernisation theory in debate: A review. *Environmental Politics, 9*(1), 17-49.
Murphy, J., & Gouldson, A. (2000). Environmental policy and industrial innovation: Integrating environment and economy through ecological modernisation. *Geoforum*, 31(1), 33-44.
Murphy, J., (2000). Ecological modernisation. *Geoforum*, 31(1), 1-8
NESREA (2021) Laws and Regulations. Available from: https://www.nesrea.gov.ng/publications-downloads/laws-regulations/ [Last accessed on 22 July 2021)

NPI. (2006). Sulphur Dioxide Fact Sheet: Department of the Environment and Water Resources, Australia. Available from: http://www.npi.gov.au. (Last accessed on 2021 January 20).

Odusanya, I.A., Adegboyega, S.B., and Kuku, M.A. (2014). Environmental quality and health care spending in Nigeria. *Fountain Journal of Management and Social Sciences*, 3(2), 57-67.

Olashore, O., (2019) Implementation of the International Legal Framework Regarding Climate Change in Developing Countries; A Review of Nigeria, Kenya, and Botswana's Environmental Provisions Governing Climate Change, *Environmental Law Review*, 21(3), 189–209.

Ole Jensen, J., & Gram-Hanssen, K. (2008). Ecological modernisation of sustainable buildings: a Danish perspective. *Building Research & Information*, *36*(2), 146-158.

Pfizer. (2020). Breakthroughs that change patients' lives. Available from https://pfe-pfizercom-d8-prod.s3.amazonaws.com/SustainabilityBondFramework.pdf (Last accessed 27 December 2021)

Pichler, P. P., Jaccard, I. S., Weisz, U., & Weisz, H. (2019). International comparison of health care carbon footprints. *Environmental Research Letters*, *14*(6), 064004.

Practice Greenhealth (2020). Seattle Children's Hospital: Employee Commuting. Available from https://practicegreenhealth.org/tools-and-resources/seattle-childrens-hospital-employee-commuting (Last accessed 18 December 2021)

Roeben, V & Azubuike, S. I. (2020). Climate Change and Responsibility: Arctic States Cooperation through the Arctic Council in Climate Change Mitigation and Adaptation Efforts. In Heininen, L & Exner-Pirot, H (eds.), *Arctic Yearbook 2020*, (443-463), Northern Research Forum, 2020.

Rufus, N. A. (2014). What is the potential for Ecological Modernisation of the Oil Industry in the Niger Delta? Master's dissertation, University of Bedfordshire.

Salas, R. N., Maibach, E., Pencheon, D., Watts, N., & Frumkin, H. (2020). A pathway to net zero emissions for healthcare. *BMJ*, 371.

Sembiring, M., (2020). Decarbonisation on a Finite Planet: A Preliminary Assessment of Environmental and Social, NTS Insight - S. Rajaratnam School of International Studies, 1–17, Available from https://www.jstor.org/stable/pdf/resrep26803.pdf (Last accessed 09 October 2021)

Spaargaren, G., & Mol, A. P. (1992). Sociology, environment, and modernity: Ecological modernisation as a theory of social change. *Society & Natural Resources*, *5*(4), 323-344.

Stringer, R. (2020). A win-win for disposing medical waste with biodigestion. Available from https://www.greengrowthknowledge.org/blog/win-win-disposing-medical-waste-biodigestion (Last accessed 20 July 2021)

Terr-Africa (2009). Land and Climate: The Role of Sustainable land Management (SLM) for Climate Change Adaptation and Mitigation in Sub-Saharan Africa (SSA). Terr Africa: Regional Sustainable Management Publication.

Toke, D. (2011). *Ecological modernisation and renewable energy*. Springer.

UNFCCC (2018) Polluted Air Affects More than 90% of Children. Available from https://unfccc.int/news/polluted-air-affects-more-than-90-of-children (Last accessed 20 July 2021)

UN Foundation & SEforALL (2019). 10 Lasting Impact: Sustainable Off-Grid Solar Delivery Models to Power Health and Education, Available from https://www.seforall.org/publications/lasting-impact-sustainable-off-grid-solar-delivery-models (Last Assessed 10 October 2021)

United Nations (2020). *Financing for Sustainable Development Report 2020*. United Nations, New York, Available from https://developmentfinance.un.org/sites/developmentfinance.un.org/files/FSDR2020_ChptIII.C.pdf. (Last accessed 14 September 2021)

Watts, N., Amann, M., Arnell, N., Ayeb-Karlsson, S., Beagley, J., Belesova, K., ... & Costello, A. (2020). The 2020 report of The Lancet Countdown on health and climate change: responding to converging crises. *The Lancet*, 397(10269), 129-170

von Malmborg, F., & Strachan, P. A. (2005). Climate policy, ecological modernization and the UK emission trading scheme. *European Environment*, 15(3), 143-160.

World Bank. (2010), The cost to Developing Countries of Adapting to Climate Change New Methods and Estimates the Global Report of the Economics of Adaptation to Climate Change Study Consultation Draft. Washington D. C.: The World Bank Group

World Bank Group. (2017). Climate-Smart Healthcare: Low-Carbon and Resilience Strategies for the Health Sector. Available from http://documents1.worldbank.org/curated/en/322251495434571418/pdf/113572-WP-PUBLIC-FINAL-WBG-Climate-smart-Healthcare-002.pdf

World Health Organisation (2015). Access to modern energy services for health facilities in resource-constrained settings: A review of status, significance, challenges, and measurement. Available from https://www.who.int/publications/i/item/9789241507646

CHAPTER 5

Optimising Hybrid Power Systems for Sustainable Operation of Remote Telecommunication Infrastructure

Michael S. Okundamiya and Samuel T. Wara

Abstract The optimal system model comprising a wind/photovoltaic hybrid power system with battery storage is designed by employing the energy-equilibrium strategy. The problem objective considered is in terms of cost, but the energy system is constrained to reliably meet power demand. The power system is evaluated for optimum size, cost, and operational efficiency utilising 22-year meteorological datasets for a case study site (latitude 11°50.9′N, longitude 13°9.6′E). The optimum size comprising a 1 kW wind turbine, 2.55 kW photovoltaic array, and a 19.36 kWh battery system can reliably and sustainably power the cellular site studied at a cost saving of 94.4%. This will enhance the African cities'

M. S. Okundamiya (✉)
Department of Electrical & Electronic Engineering, Ambrose Alli University, Ekpoma, Nigeria
e-mail: msokundamiya@aauekpoma.edu.ng

S. T. Wara
Havilla University, Nde-Ikom, Nigeria

S. I. Azubuike et al. (eds.), *Decarbonisation Pathways for African Cities*, Palgrave Studies in Climate Resilient Societies,
https://doi.org/10.1007/978-3-031-14006-8_5

decarbonisation agenda by switching from conventional fossil fuel to a carbon-neutral energy system to benefit the immediate operational environment and the city at large.

Keywords Cellular generation sites • Decarbonisation • Energy optimisation • Hybrid energy system • Renewable energy harvesting • Wind turbine generator

5.1 Introduction

5.1.1 Background

The world electricity demand is rapidly increasing. The projected growth of about 50% is expected in the next three decades (Okundamiya 2021a). Africa's electricity demand is anticipated to increase threefold by 2030 (An African Energy Industry Report 2018). Unfortunately, according to a recent report (International Energy Agency 2021), 75% of the global population without access to electrical energy live in sub-Saharan Africa. More than 80 million persons are unable to access grid electricity in Nigeria, the most densely inhabited African nation. Regardless of this hitch, the electric power grid is not reliable (Okundamiya et al. 2022). The growing demand for a sustainable energy system has made alternative power sources a promising field of investigation (Ghenai and Bettayeb 2019; Sun and Leto 2020). The worldwide trend has enhanced the extensive use of non-diminishing energy resources with negligible ecological effects (Okundamiya et al. 2014a; Tang et al. 2018; Bukar and Tan 2019; Okundamiya and Omorogiuwa 2016).

Renewable sources are progressively becoming competitive in the global energy market (IRENA 2018). The last decade witnessed impressive advancement in the competitiveness of wind and solar power technologies. The advances have been propelled by competitive supply chains, economies of scale, steadily expanding technologies, and advancing developer experience. The cost of electricity from offshore wind, onshore wind, and utility-scale solar photovoltaic (PV) dropped 48%, 56%, and 85%, respectively. The trend shows that renewables are not only contesting fossil fuels but substantially underselling them (IRENA 2021). The cost declines are significant, providing massive upshots for the attractiveness of renewable energy technologies over the medium term, and making wind and solar power generation technologies the economic backbone of the energy transition. In addition, they offer an understanding of the features of technologies which are responsive to speedy upgrades and cost savings to ensure the decarbonisation of end-use sectors.

As wind and solar power costs have fallen, capacity additions have grown, reducing yearly fossil combustion in several countries worldwide, including India and the United States. Reliable and enhanced wind turbines, along with larger rotor diameters and higher hub heights have enhanced capacity factors. Also, total installed costs along with the operation and maintenance (O&M) costs have been dropping due to economies of scale, better efficiency, and development of the sector. Turbines with larger rotor diameters increase energy capture at sites with the same wind speed, whereas higher hub heights are enabling higher wind speeds to be accessed at the same location, while also increasing the range of suitable locations for wind turbines. Moreover, turbine prices have declined in a stepwise fashion to an average of USD540/kW in 2020 (IRENA 2021). Nevertheless, renewable resources are site-dependent and intermittent. Therefore, storage systems need to be integrated into the system design to enhance the power system stability (Okundamiya et al. 2014b, 2017; Okundamiya and Nzeako 2010).

Hybrid power systems are now extensively utilised as they intermix different power resources to supplement the shortfalls of each energy resource (Okundamiya 2021a; Bukar and Tan 2019). Although wind and solar power systems are commonly used for energy harvesting in domestic and commercial applications due to proven sustainability with negligible carbon emission, the viability of hybrid power systems depends on a variety of parameters including the mix of energy resources, distributed capacity, and energy in the dispatching scheme (Okundamiya and Omorogiuwa 2016). The hybrid power system is capable of supplying the required energy in cellular generation sites, but crucial concerns such as reliability and the enabling technology for power system stability must be resolved. It is essential to note that renewable technologies require higher initial capital compared to fossil technology. Consequently, investors are cautious when evaluating their options given the diversity of design architecture to sustain the rising capacity projections at optimal cost cleanly and sustainably (Okundamiya et al. 2014a). The strategic performance indices that can inform an investor's choice of the most appropriate design are scarce in the literature. Also, hybrid power solutions are rarely utilised in cellular mobile sites. This work is intended to fill this gap in the literature. Given the proliferation of such sites within and around African cities, any measures to substitute their fossil fuel-based power supply with cost-effective and sustainable hybrid power systems will contribute to the decarbonisation agenda of such cities.

5.1.2 *Cellular Mobile Generation Sites*

Cellular mobile technology has witnessed tremendous growth in recent times. Nigeria has one of the biggest mobile telecommunication markets in Africa; her teledensity stood at 102.88% with a subscriber base of 196,379,542 as of June 2020 (Nigerian Communications Commission 2020). The speedy evolution of mobile telecommunication creates some drawbacks such as poor service delivery and network congestion (Okundamiya et al. 2017). The operating firms are not able to rapidly extend their core network coverage to meet the rising demands of cellular mobile services across the country due to the dearth of electricity. Also, the number of cell sites increases with each cellular generation technology to sustain the rising rates of call and data traffic. This trend is deeply intensified by the launch of the fifth-generation (5G) technology because of the widespread multiplicity of use cases as well as traffic patterns that need to be sustained. The high demand for mobile data has motivated an enormous deployment of cellular generation infrastructure (Yan et al. 2019). Table 5.1 shows the evolution of site topology of cellular generation. The current technology offers ultra-high bandwidth with ultra-low latency, as well as vast connection capabilities, to allow for diversified services

Table 5.1 Evolution of site topology of cellular generation

Generation era	*First-generation (1G)*	*Second generation (2G)*	*Third generation (3G)*	*Fourth generation (4G)*	*Fifth-generation (5G)*
Site topology (Programmersought 2021)	Base Station (BS)	Base Transceiver Station (BTS)	Node B (NB)	Evolved Node B (eNB)	next Generation Node B (gNB)
Technology = Internet service	AMPS	GSM Narrowband	UMTS Broadband	LTE Ultra-broadband	MIMO Wireless World Wide Web

AMPS Advanced Mobile Phone Service, *GSM* Global system for mobile communication, *MIMO* Multiple-input multiple-output, *LTE* Long-term evolution, *UMTS* Universal mobile telecommunications system anywhere at any time. This is intended to meet the ever-growing data consumption requirements due to technological enhancements (GSMA Intelligence).

A typical cellular generation site comprises a tower with antennas and a lighting system, and a shelter. The site can either exist on the periphery (only permits transport network for the cell) or is positioned on the main network to enable transport network for several cells. The tower height of a macro-cell site with 1–35 km coverage used in remote areas is higher (34–55 m) than that of a micro-cell site (25 m) with coverage of 0.1–1 km used for cities (Okundamiya et al. 2022). Since 5G relies on a large number of spectrum bands, the type and number of antennas per site will increase. Consequently, the use of a large number of antennas and cells in addition to the critical infrastructure monitoring of the technology requires an efficient power management system with a very high level of reliability (zero downtime) to optimise the overall power consumption (Gabriel 2019).

Although, the total power requirement for a 5G site is more than three times a 4G site's power requirement and the total power consumption for the former can be 12 times more than that of the latter if considered for high-density deployment (Chih-Lin et al. 2020). Nevertheless, the energy consumption per unit of data (watt/bit) is much less, with higher throughput to serve more users simultaneously, for 5G than 4G. The substantial increase in power consumption of the present cellular generation technology with a higher density of infrastructure requires sustainable energy technology for the reliable operation of the site equipment. This will help to optimise the energy efficiency of the network, without negotiating the spectrum efficiency. Consequently, modelling and optimising the power system with hybrid renewable power resources, to reliably and cost-effectively power cellular mobile sites cleanly and sustainably is a viable solution to the present electricity problem. As a consequence, the immediate operational environment of the cell site and the city at large can be effectively decarbonised.

5.2 Methodology

The architecture and model of the hybrid power system (HPS) for off-grid cellular sites, as shown in Fig. 5.1, comprise a PV system, a wind energy system, a battery bank, and the power electronic converters. The energy system is interconnected to the diesel generator system (used for comparison only) using a power converter.

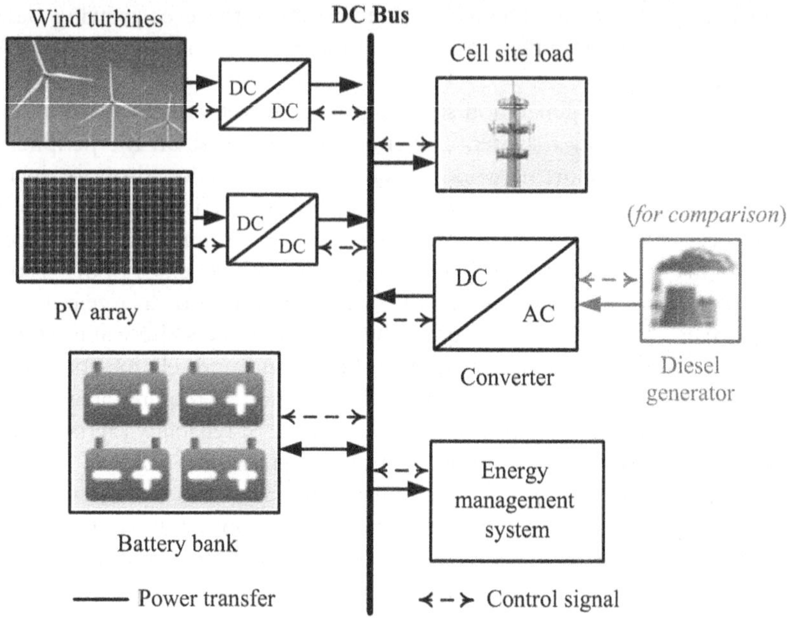

Fig. 5.1 Architecture of HPS for off-grid cellular sites

The hybrid optimisation model for electric renewables (HOMER) software tool is used for constituent sizing as it permits assessment with diverse design strategies (Okundamiya and Omorogiuwa 2015).

5.2.1 Wind Energy System

There are several models employed in the literature to describe the performance of wind turbines (WTs) (Kusiak and Song 2010; Okundamiya and Ojieabu 2017; Okundamiya 2016; Okundamiya and Nzeako 2013). Notwithstanding, designing a WT model requires the consideration of several parameters, which affect the WT output. The first is the inclusion of the random characteristics of the site resource, in a suitable model to replicate the sequential properties of the wind speed (Billinton and Gao 2008). The next is the correlation between wind resources and power output. The correlation can be established using the WT-specified operating parameters. The constraints normally used are the cut-in, the rated, and the cut-out wind speeds (Okundamiya and Nzeako 2013). The third

Table 5.2 The techno-economic description of the chosen WT (IRENA 2021; https://www.windpowercn.com/products/31.html)

Parameter	*Specifications*
Model	SWT-1kW
Manufacturer	Senwei Energy Technologies Inc
Rotor diameter (m)	2.4
Rated/Maximum power output (kW)	1/1.05
Operating voltage (VDC)	24/48
Operating wind speed (m/s)	3-20
Cut-in/Nominal wind speed (m/s)	2.0/11.0
Storm-stand (m/s)	Up to 40
Maximum rotational speed (r/min)	450
Working temperature range (°C)	-40 to +60
Hub height (m)	10
Speed regulation method	Yawing and electric magnet dumping
Blades quantity/type/material	3 Blades/Upwind/Glass fibre reinforced plastic
Lifespan (y)	20
Installation cost ($/kW)	1500
Replacement cost ($/kW)	1100
O&M cost ($/kW/y)	30

Note: $1 ≈ ₦410.35 (Central Bank of Nigeria, accessed September 1, 2021)

O&M operation and maintenance

is the unavailability of the WT. The unavailability component typically accounted for using reliability indices, is important since the WT power output can be unavailable, perhaps because of maintenance. The techno-economic description of the chosen WT is given in Table 5.2.

5.2.2 *Solar Photovoltaic System*

The output power of a photovoltaic system is computed using the equation (Okundamiya and Omorogiuwa 2015):

$$P_{PV} = f_{PV} Y_{PV} \left(I_T / I_S \right), \tag{5.1}$$

f_{PV} is the PV de-rating fraction, Y_{PV} (kW) is the rated size of the array, I_T (kW/m^2) is the global solar irradiation incident on the surface of the array, and I_S (kW/m^2) is the standard amount of irradiation used to rate the power of the photovoltaic system, typically given as 1 kW/m^2. The technical and economic description of the selected PV module has been examined in the literature (Okundamiya 2021b); the installation cost is \$500/kW, the replacement cost is \$450/kW, and the O&M cost is \$1/kW/y.

5.2.3 *Energy Storage System*

Due to the intermittence of the renewable power source, an energy storage scheme is required to balance the gap in energy demand. Electrochemical batteries are currently the most widely used renewable energy storage systems due to their relatively high safety level, high recyclability, and low cost (Krishan and Suhag 2020). The capacity of battery storage, commonly represented as C in kilowatt-hour (kWh) and ampere-hour (Ah) is essential to resolve the energy system reliability. The C_{kWh} of the selected battery is stated in Eq. (5.2) (Olatomiwa et al. 2015).

$$C_{kWh} = \left(bA_d E_L \times DOD\right)/1000 \tag{5.2}$$

where, η_b is the battery efficiency, A_d is the daily autonomy, E_L is the mean daily load energy (kWh/d), and DOD is the depth of discharge of the battery bank.

The battery size was established using the auto-size mode in HOMER. To secure greater voltage and current rating, series and parallel arrangements of the selected batteries (Table 5.3) are employed.

Table 5.3 Techno-economic description of the selected battery (IRENA 2021; https://www.bae-berlin.de/)

Parameter	*Specifications*
Model type	BAE PVV210
Capacity Ratio	0.375
Constant Rate (h^{-1})	1.06
Lifetime Throughput (kWh)	2112.40
Maximum Capacity (Ah)	201
Maximum Charge Current (A)	33.4
Maximum Charge Rate (A/Ah)	1
Maximum Discharge Current (A)	361
Nominal Capacity (kWh)	2.42
Nominal Voltage (V)	12
Round trip efficiency (%)	90
Capital cost (\$)	310
Replacement cost (\$)	300
O&M cost (\$/y)	10

5.2.4 *Power Converter and Diesel Generator*

A power converter allows the transfer of electric power, which permits the production of appropriate power magnitude between the power resources and the allocation for continuously satisfying the power demand (Okundamiya et al. 2014a). The main concept is to combine the hybrid energy system, to reliably and cost-effectively power cellular mobile sites cleanly and sustainably. In this study, the power converter size was established using the auto-size mode in HOMER. Details of the converter and generator (for comparison) system employed are shown in Table 5.4.

5.2.5 *Study Site, Data Collection, and Analysis*

A description of the meteorology of the site under study (latitude 11°50.9′N, longitude 13°9.6′E) can be found in (Okundamiya et al. 2022). The meteorological datasets used for analysis were gathered from different archives. The first part comprises daily averages of ground-based wind resources retrieved from Nigerian Meteorological agency (NIMET), Oshodi, Nigeria. The data was derived from observation during a 22-year cycle (1991–2012). The data were adjusted to the wind turbine hub height of 30 m using power law, a widely used method to model wind

Table 5.4 Description of converter/generator used in the hybrid system design (Okundamiya 2021a; Ghenai and Bettayeb 2019)

Component	*Parameter*	*Specification*
Converter	Capital cost ($/kW)	100
	O&M cost ($/kW/y)	0.5
	Lifespan (y)	25
	Efficiency (%)	95
	Search space (kW)	0–50
Controller	Capital cost ($)	100
	O&M cost ($/kW/y)	1
	Lifespan (y)	25
Diesel generator (DG)	Capital cost ($/kW)	370
	Replacement cost ($/kW)	300
	O&M cost ($/kW/h)	0.15
	Capacity (kW)	25
	Lifespan (h)	15,000
	Fuel cost ($/L)	0.61

Table 5.5 Monthly averages for meteorological data for Maiduguri (Okundamiya et al. 2014b; NASA (National Aeronautics and Space Administration), Prediction of Worldwide Energy Resource (POWER) database)

Months	*Wind speed (m/s)*	*Daily radiation (kWh/m²/day)*	*Clearness index*
January	7.50	5.610	0.649
February	8.80	6.300	0.671
March	8.80	6.700	0.661
April	6.70	6.620	0.628
May	7.00	6.360	0.604
June	6.50	5.970	0.573
July	8.20	5.430	0.520
August	7.00	5.140	0.491
September	5.80	5.570	0.545
October	5.60	5.890	0.616
November	7.00	5.840	0.665
December	7.60	5.350	0.639

shear (Okundamiya and Omorogiuwa 2016). The second dataset, retrieved from the National Aeronautics and Space Administration, comprises monthly averages for global horizontal radiation from July 1984 to June 2005 (NASA (National Aeronautics and Space Administration), Prediction of Worldwide Energy Resource (POWER) database). Table 5.5 displays the monthly averages for wind and solar resources used while Table 5.6 describes the main characteristics of the wind data.

The observed daily load of a 4G 3-sector (LTE eNode B 2/2/2) cellular generation site (Fig. 5.2a) was used as a case study. The simulated load profiles of the 4G site (Fig. 5.2b, c) with a peak load of 1.47 kW and an annual average of 11.30 kWh/d were deduced from the observed load profile using the HOMER tool (Version 3.14.4) by taking daily variations of 15% and hourly variations of 20%. An inflation rate (13.5%), discount rate (8%), and lifecycle (25y) are projected for this study (Okundamiya 2021a; Macrotrends).

5.3 Results and Discussion

5.3.1 Optimised WT/PV/Battery Hybrid Power System

The simulation process, generated using HOMER Pro, allows for optimisation and sensitivity assessment of the model, to establish the least-cost, most eco-friendly, and most effective risk-mitigation design strategy, in terms of

Table 5.6 Characteristics of meteorological datasets collected from NIMET (Okundamiya et al. 2014b)

Instrument	*Unit*	*Measurement range*	*Recording resolution*	*Accuracy*
Wind vane (for direction)	degrees (°)	0–360	≤1	±2
Anemometer (for speed)	m/s	0.30–50	≤0.10	±0.30

the aggregate net present cost (NPC), as well as COE of the energy system, and subsequently assessed with the frequently utilised DG/battery hybrid power system. The optimal size of the hybrid power system comprising a 1 kW wind turbine, 2.55 kW PV system, and 19.36 kWh battery bank (2 strings of 4 batteries each) can reliably sustainably power the cellular mobile site at a total cost of $0.06343 per kWh. The monetary cost details of the subsystem constituents are shown in Table 5.7 while the effect of the individual subsystem on the overall cost is presented in Fig. 5.3. As observed, the battery (energy storage) subsystem tops the entire subsystem's cost (65%), followed by the WT subsystem (24%), then PV (10%).

Figure 5.4 shows the annual electric power output of the subsystem generators. The wind energy system has a mean output power of 0.399 kW during 8746 hours of operation with a capacity factor of 39.9%, wind penetration of 84.7%, and maximum output power of 1.05 kW at a levelised cost (LCOE) of $0.0177 per kWh (Fig. 5.4a). The PV system has a mean daily energy output of 13.3 kWh during the operation time of 4362 h with a capacity factor of 21.8%, a maximum power output of 2.65 kW, and a PV penetration of 118%, at an LCOE of $0.00569 per kWh (Fig. 5.4b). It is worthy to mention that although the PV subsystem accounts for 10% of the overall system cost, it produces 58.2% (4859 kWh/y) of the entire electricity generation compared to the WT subsystem's 41.8% (3492 kWh/y). This suggests that the PV system is more viable in the study area compared to other options.

Figure 5.5 presents the state-of-charge (SOC) of the battery system. The storage system comprising 8 batteries arranged in 2 strings of 4 batteries (connected in series) provides an autonomy of 28.7 h at a storage wear cost of $0.15 per kWh with a usable nominal capacity of 13.5 kWh, the annual throughput of 1099 kWh/y, storage depletion of 1.33 kWh/y, and energy losses of 116 kWh/y, whereas the cumulative energy inflow is 1157 kWh/y with an outflow of 1043 kWh/y.

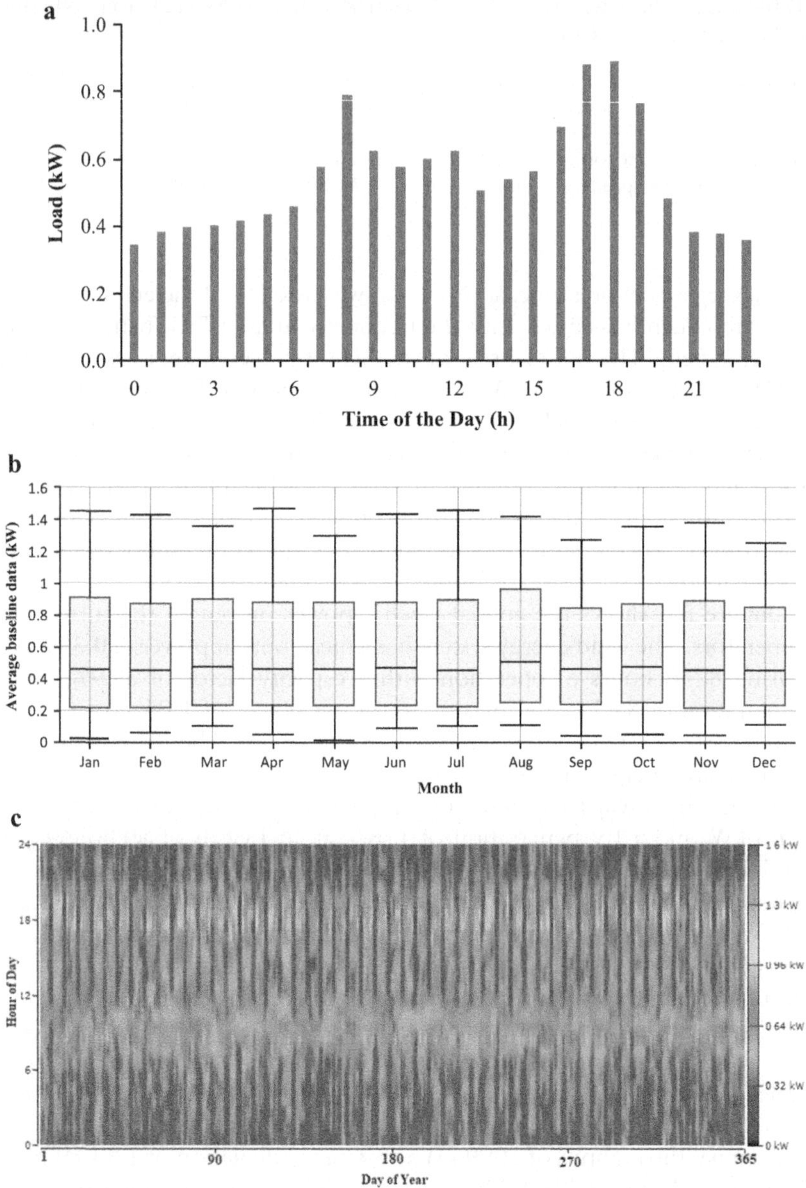

Fig. 5.2 Load profiles of the 4G LTE eNode B 2/2/2 infrastructure under study

Table 5.7 Monetary cost details of the subsystem constituents for proposed WT/PV/battery hybrid power system for4G LTE eNode B 2/2/2 Cellular Site

	Capital ($)	*Replacement ($)*	*O&M ($)*	*Salvage ($)*	*Total ($)*
WT	1500.00	2970.54	1524.08	(2855.99)	3138.63
PV	1275.00	0.00	129.55	0.00	1404.55
Battery	2480.00	5151.62	4064.22	(3109.68)	8586.15
Controller	100.00	0.00	50.80	0.00	150.80
Hybrid system	5355.00	8122.16	5768.65	(5965.67)	13,280.13

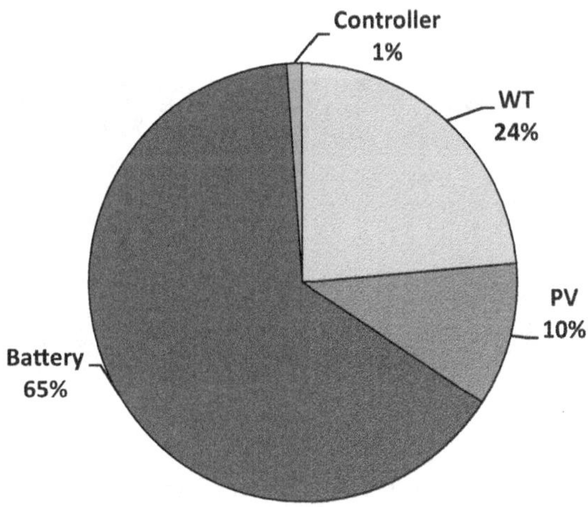

Fig. 5.3 Effect of each subsystem on the total cost of WT/PV/Battery Hybrid Power System

5.3.2 *Techno-economic Comparison*

Table 5.8 compares the techno-economy of proposed and baseline hybrid systems for the 4G cellular infrastructure under study. In comparison with the baseline DG power source, the annual power generation from the hybrid (WT and PV) generators greatly exceeds the electric power demand because hybrid energy systems with renewable resources generate surplus electricity sporadically to sustain a continuous power supply. Besides, the operating power reserve (i.e., 20% assumed in this study) is essential to prevent power system failure in the occurrence of a sudden rise in energy demand (Okundamiya 2021a). The 49.3% difference in electricity generation (i.e., 4115 kWh/y) of the hybrid energy system can be used to meet

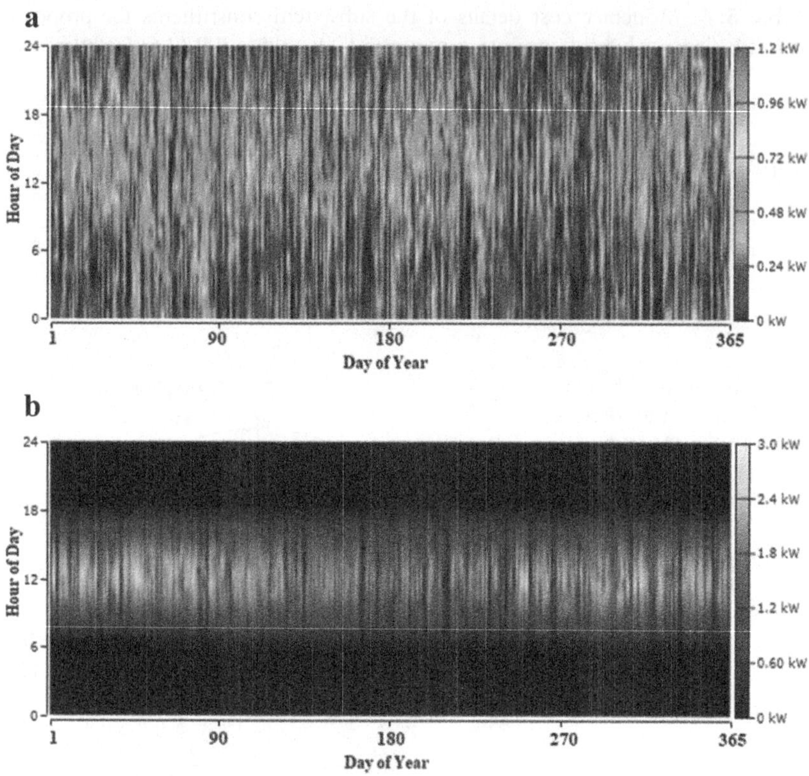

Fig. 5.4 Annual electric power output of subsystem generators (**a**) WT (**b**) PV

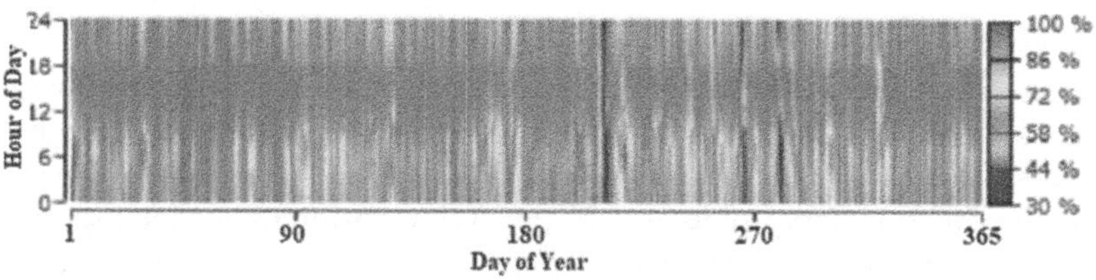

Fig. 5.5 SOC of the battery system

Table 5.8 Comparison of techno-economy of WT/PV/Battery (Proposed) and DG/Battery/Converter (Baseline) Hybrid Systems

Techno-economy	*Baseline system*	*Proposed system*
Energy generation (kWh/y)	5120	8351
Energy consumption (kWh/y)	4124	4121
Excess electricity (%)	7.78	49.3
Capacity shortfall (%)	0.0	0.138
Unmet load (%)	0.0	0.0785
Renewable portion (%)	0.0	100
Pollutant emissions		
Carbon dioxide (kg/y)	5183	0.0
Carbon monoxide (kg/y)	32.4	0.0
Unburned hydrocarbons (kg/y)	1.43	0.0
Particulate matter (kg/y)	0.194	0.0
Sulphur dioxide (kg/y)	12.7	0.0
Nitrogen oxides (kg/y)	30.4	0.0
NPC ($)	235,903	13,280
COE ($/kWh)	1.13	0.06343
Initial capital ($)	7890	5355
Operating cost ($/y)	4488	156

the power demand of the neighbouring centres since the current legislation in the study area does not encourage electricity sales to the electric grid.

Table 5.8 suggests that usage of the proposed hybrid energy option can mitigate the yearly pollutant emissions within the range of 0.194–5183 kg/y. In particular, the implementation of the proposed WT/PV/Battery hybrid power system for 4G cellular sites in Maiduguri, Nigeria can attain an annual CO_2 cut of 5183 kg/y, CO (32.4 kg/y), nitrogen oxides (30.4 kg/y), sulphur dioxide (12.7 kg/y) whereas the unburned hydrocarbon and particulate matter can be as low as 1.43 kg/y and 0.194 kg/y, respectively. This can make the environment more eco-friendly and cleaner. It is imperative to state that the zero-carbon emission value of the proposed hybrid energy system suggests that it is capable of decarbonising the immediate surroundings.

The economy of an energy system can be assessed using the NPC and the COE of the system (Okundamiya 2021b). The practicable design strategy with the lowest COE and NPC is considered the most viable strategy. As seen in Table 5.8, the proposed hybrid power system with battery storage reduced the NPC of the DG/Battery (baseline) system by

94.4% (from $235,903 to $13,280) with a resultant saving in the COE (from 1.13$/kWh to 0.06343$/kWh). Unlike the conventional DG-only power system with low initial capital, the high initial capital of the baseline system ($7890 compared to $5355) is attributed to the high capital cost of the integrated battery bank (presently used to relieve the DG from continuously operating) noting that the costs of renewables are fast declining. In addition, the baseline DG/Battery system operation is not yet optimised. Consequently, the present worth of $222,623 can be recovered by installing the proposed energy system with battery storage as an alternative to the DG/Battery system for 4G cellular sites in Maiduguri, Nigeria. Figure 5.6 compares the cumulative cash flows of the proposed and baseline systems over the project lifespan.

5.3.3 *Impact of Inflation on Hybrid System Techno-economy*

Although renewable sources are progressively becoming more competitive in the global market over the last decade, Nigeria's inflation index has not been stable. The inflation rate has grown from 13.72% in 2010 to 17.38% in August 2021; with a low of 8.06% in 2016, and an average of 12.73% (Macrotrends; https://tradingeconomics.com/nigeria/inflation-cpi). Figure 5.7 shows the influence of rising inflation on the economy of the optimised WT/PV/Battery system at a discount rate of 8%.

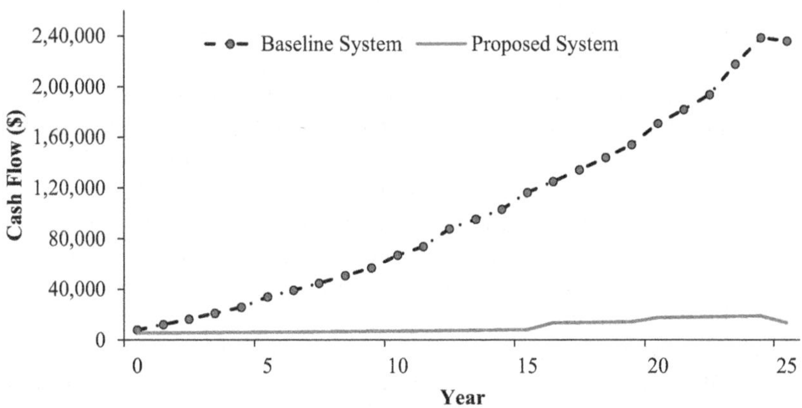

Fig. 5.6 Comparison of cumulative cash flow of the baseline and proposed systems over the project lifespan

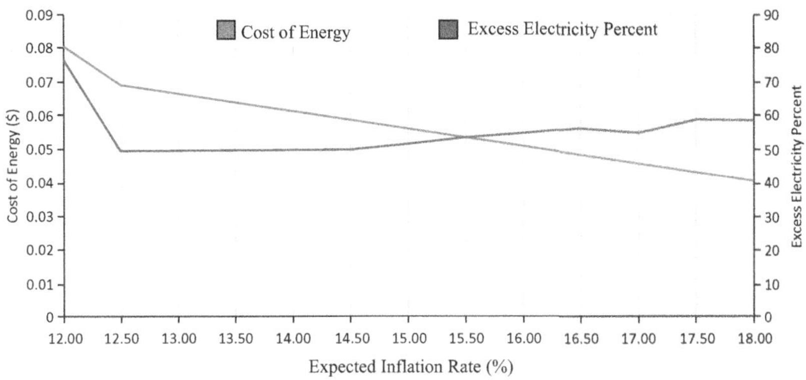

Fig. 5.7 Influence of inflation on the economy of the optimised WT/PV/Battery system at a discount rate of 8%

As noticed (Fig. 5.7), the increasing inflation ratio enhances the hybrid energy system's monetary cost (COE) per kWh. The result suggests that the installation of the proposed hybrid system at a present inflation rate of over 17% can enable sustainable harvesting of electricity for the 4G infrastructure understudy at a cost of approximately \$0.043 per kWh. Conversely, the percentage excess electricity generation rises beyond a turning point (inflation rate) of 12.5% indicating the enormous viability of the energy pathway amid rising inflation rates. This study recommends the integration of a hydrogen fuel cell (HFC) to the proposed (WT/PV/Battery) hybrid power since current legislation in the area under study does not support the sale of excess electricity generation to the utility grid. The inclusion of the HFC can enable the optimal harvest of the excess energy to be stored as hydrogen fuel that will be used to power the fuel cell (FC) system during periods of deficit wind/solar generation.

5.4 Conclusion

An optimised mixed energy system, to reliably and cost-effectively power cellular mobile sites cleanly and sustainably, has been modelled and presented. The focus was to evaluate the viability of the shift from the commonly used diesel/battery power option to a more sustainable energy system pathway. The optimal design model comprising hybrid wind/photovoltaic power conversion and battery storage systems was confirmed by employing the energy-equilibrium procedures of HOMER. The problem objective considered is in terms of cost, but the energy system was constrained to reliably meet the power demand. Evaluation of the optimum sizing, cost, and operation of the power system was carried out using 22-year datasets for the study site in Nigeria. The result showed that a hybrid system containing a 1 kW wind turbine, 2.55 kW photovoltaic array, with 19.36 kWh battery storage system can reliably power the 4G cellular mobile site understudy sustainably at a COE of $0.06343 per kWh.

The contribution of this study centres on proposing an optimum design configuration for a 4G cellular mobile site that will efficiently replace the existing diesel/battery option for a cleaner and more sustainable power production. Moreover, the implementation will:

a) enhance the quality of cellular mobile systems by making their operation more reliable;
b) lower the per-unit cost of cellular telecommunication services;
c) slash CO_2 emissions of cell sites, which could make the ecosystem more friendly and safe;
d) assist in developing strategies for reliable and sustainable energy supply;
e) enhance capacity building and training of energy professionals and various stakeholders in understanding how existing constraints of remote applications which depend on fossil fuels can be suitably reconfigured to achieve cost efficiency/effectiveness and sustainability; and
f) provide the impetus for further research for the deployment of the concept to other areas of application, where remote power usage needs to be optimised for cost efficiency/effectiveness and sustainability.

Given the projected growth of the telecommunication sector and similar applications characterised by remote power/electricity requirements in African cities and settlements, the benefits of switching fuel sources to reduce the carbon footprint within such cities and their environment are enormous. The decarbonisation agenda for African cities would entail perfecting the design and optimisation of such systems and/or their application, as well as making policy and regulatory changes that would increase their uptake. For instance, schemes like feed-in tariffs that make it easier for excess electricity to be fed into the grid will incentivise private sector investment in renewable energy.

List of Abbreviations

1G	First generation
2G	Second generation
3G	Third generation
4G	Fourth generation
5G	Fifth generation
AMPS	Advanced Mobile Phone Service
BS	Base Station
BTS	Base Transceiver Station
eNB	Evolved Node B
FC	Fuel cell
gNB	next Generation Node B
GSM	Global system for mobile communication
HOMER	Hybrid optimisation model for electric renewables
LTE	Long-term evolution
MIMO	Multiple-input multiple-output
NB	Node B
NIMET	Nigerian Meteorological agency
NR	New radio
PV	Solar photovoltaic
SDGs	Sustainable Development Goals
UMTS	Universal mobile telecommunications system
WT	Wind turbine

References

M. S. Okundamiya, “Size optimisation of a hybrid photovoltaic/fuel cell grid connected power system including hydrogen storage,” International Journal of Hydrogen Energy, vol. 46, no. 59, pp. 30539-30546, 2021a.

An African Energy Industry Report, “Market Intelligence Report 2018,” ISPY publishing limited, pp. 1-8, May 2018.

International Energy Agency “World-energy-outlook-2020,” World Energy Outlook. https://www.iea.org/reports/world-energy-outlook-2020 [Jan. 23, 2021].

M. S. Okundamiya, S. T. Wara, H. I. Obakhena, “Optimization and techno-economic analysis of a mixed power system for sustainable operation of cellular sites in 5G era,” International Journal of Hydrogen Energy, vol. 47, no. 39, pp. 17351-17366, 2022.

C. Ghenai, M. Bettayeb, "Modelling and performance analysis of a stand-alone hybrid solar PV/fuel cell/diesel generator power system for university building," Energy, vol. 171, pp. 180-189, 2019.

C. Sun, S. Leto, "A novel joint bidding technique for fuel cell wind turbine photovoltaic storage unit and demand response considering prediction models analysis effect's," International Journal of Hydrogen Energy, vol. 45, pp. 6823-6837, 2020.

M. S. Okundamiya, J. O. Emagbetere, E. A. Ogujor, "Design and control strategy for a hybrid green energy system for mobile telecommunication sites," Journal of Power Sources, vol. 257, pp. 335-343, 2014a.

S. Tang, H. Honga, J. Sunc, W. Qu, "Efficient path of distributed solar energy system synergetically combining photovoltaics with solar-syngas fuel cell," Energy Conversion Management, vol. 173, pp. 704–714, 2018.

A. Bukar, C. W. Tan, "A review on stand-alone photovoltaic-wind energy system with fuel cell: System optimisation and energy management strategy," Journal of Cleaner Production, vol. 221, pp. 73-88, 2019.

M. S. Okundamiya, O. Omorogiuwa, "Analysis of an isolated micro-grid for Nigerian terrain," 59th Int Midwest Symposium on Circuits & System Abu Dhabi, UAE, 2016.

IRENA, *Global Energy Transformation: A roadmap to 2050*, International Renewable Energy Agency, Abu Dhabi, 2018.

IRENA, *Renewable Power Generation Costs in 2020*, International Renewable Energy Agency, Abu Dhabi, 2021.

M. S. Okundamiya, V. O. A. Akpaida, B. E. Omatahunde, "Optimisation of a hybrid energy system for reliable operation of automated teller machines," Journal of Emerging Trends Engineering & Applied Sciences, vol. 5, no. 8, pp. 153-158, 2014b.

M. S. Okundamiya, J. O. Emagbetere, E. A. Ogujor, "Modelling and optimum capacity allocation of micro-grids considering economy and reliability," Journal of Telecommunication, Electronic and Computer Engineering, vol. 9, no. 4, pp. 55-61, 2017.

M. S. Okundamiya, A. N Nzeako, "Energy storage models for optimising renewable power applications," Journal of Electrical Power Engineering, vol. 4, no. 2, pp. 54-65, 2010.

Nigerian Communications Commission, Retrieved https://www.ncc.gov.ng/statistics-reports/subscriber-data [Aug. 4, 2020].

M. Yan, C. A. Chan, A. F. Gygax, J. Yan, L. Campbell, A. Nirmalathas, and C. Leckie, "Modelling the total energy consumption of mobile network services and applications," Energies, vol. 12, no. 184, 2019, https://doi.org/10.3390/en12010184

GSMA, Network Experience Evolution to 5G, United Kingdom: GSMA Intelligence, www.gsma.com

Programmersought, The history of the base station, NR 1G-5G base station introduction, https://www.programmersought.com/article/18414157411/ [Jan. 22, 2021]

C. Gabriel, "What are key considerations for 5G sites?" Analysys Mason Limited, September 2019

I. Chih-Lin, H. Shuangfeng and B. Sen, "Energy-efficient 5G for a greener future," Nature Electronics, vol. 3, pp. 182–184, 2020.

M. S. Okundamiya, O. Omorogiuwa, "Viability of a photovoltaic diesel battery hybrid power system in Nigeria," Iranica Journal of Energy and Environment, vol. 6, no. 1, pp. 5-12, 2015.

A. Kusiak, Z. Song, "Design of wind farm layout for maximum wind energy capture," Renewable Energy, vol. 35, no. 3, pp. 685–694, 2010.

M. S. Okundamiya, C. E. Ojieabu, "Optimum design, simulation and performance analysis of a micro-power system for electricity supply to remote sites," Journal of Communications Technology, Electronics and Computer Science, vol. 12, pp. 6–12, 2017.

M. S. Okundamiya, "Power electronics for grid integration of wind power generation system," Journal of Communications Technology, Electronics and Computer Science, vol. 9, pp. 10-16, 2016.

M. S. Okundamiya, A. N. Nzeako, "Model for optimal sizing of a wind energy conversion system for green-mobile applications," International Journal of Green Energy, vol. 10, pp. 205–218, 2013.

R. Billinton and Y. Gao, "Multistate wind energy conversion system models for adequacy assessment of generating systems incorporating wind energy," *IEEE Transactions on Energy Conversion*, vol. 23, no. 1, pp. 163-170, 2008.

M. S. Okundamiya. "Integration of photovoltaic and hydrogen fuel cell system for sustainable energy harvesting of a university ICT infrastructure with an irregular electric grid", Energy Conversion and Management, vol. 250, no. 114928, 2021b, https://doi.org/10.1016/j.enconman.2021.114928

O. Krishan, S. Suhag, "Grid-independent PV system hybridisation with fuel cell-battery/ super capacitor: Optimum sizing and comparative techno-economic analysis," Sustainable Energy Technologies and Assessments, vol. 37, no. 100625, 2020; https://doi.org/10.1016/j.seta.2019.100625

L. Olatomiwa, S. Mekhilef, A. S. N. Huda, K. Sanusi, "Techno-economic analysis of hybrid PV-diesel-battery and PV-wind-diesel-battery power systems for mobile BTS: the way forward for rural development," Energy Sci Eng, vol. 3, pp. 271–85, 2015.

NASA (National Aeronautics and Space Administration), Prediction of Worldwide Energy Resource (POWER) database, Monthly averages for global horizontal radiation over 22-year period (from July 1983 – June 2005), available online http://eosweb.larc.nasa.gov/; (accessed on January 31, 2021).

Macrotrends, Nigeria Inflation Rate 1960-2021 (Online), https://www.macrotrends.net/countries/NGA/nigeria/inflation-rate-cpi, Retrieved 2021-09-01

CHAPTER 6

Performance Analysis of a Grid-Linked Microgrid System in a University Campus

Samuel T. Wara and Michael S. Okundamiya

Abstract A grid-linked photovoltaic/fuel cell microgrid optimised for a Nigerian university building is analysed. The objective is to establish the techno-economic and environmental merits of integrating renewable power resources into an erratic electric power grid. The energy-balance method was used for the design. Case study simulations were carried out for an electric load at Ambrose Alli University campus (latitude 6°44.0′N, longitude 6°5.1′E) using 22-year period datasets gathered from the National Aeronautics and Space Administration. The results show that hybridisation of the photovoltaic generator (340 kW), power converter (80 kW), fuel cell system (22 kW), electrolyser (21 kW), and hydrogen tank (28 kg) can sustainably augment the existing

---

S. T. Wara (✉)
Havilla University, Nde-Ikom, Nigeria
e-mail: stwara@havillauniversity.edu.ng

M. S. Okundamiya
Department of Electrical & Electronic Engineering, Ambrose Alli University, Ekpoma, Nigeria

S. I. Azubuike et al. (eds.), *Decarbonisation Pathways for African Cities*, Palgrave Studies in Climate Resilient Societies,
https://doi.org/10.1007/978-3-031-14006-8_6

grid-diesel generator system. The potential deployment of this design on university campuses in African cities would represent great strides towards decarbonisation.

Keywords Decarbonisation • Energy optimisation • Fuel cell • Hydrogen storage • Renewable energy harvesting • Unreliable electric grid

6.1 Introduction

The worldwide demand for electric power is increasing due to rapidly growing economies and populations. The growth in energy demand has led to insufficient access to electricity; the rural and emerging communities suffer mainly from this severe shortfall (Okundamiya 2021a). One of the barriers to economic development in sub-Saharan Africa (SSA) is the lack of sustainable electrical power (Okundamiya 2021b). About 75% of the global populace with no access to electrical energy live in SSA; over 80% of these people are rural inhabitants with an electrification rate below 25% compared with 71% in urban areas (Energy Access Outlook 2017). SSA has diverse potential for rapid economic growth. Yet, the major source of electrification in the region is grid extension, which is generally limited by installation costs, an unstable investment climate, and difficult geography. The accessibility of sustainable energy can facilitate the transformation of an emerging economy.

Renewable sources are progressively becoming competitive in the global energy market (IRENA 2018). The rising demand for a sustainable power system is driving the widespread use of non-diminishing energy resources with negligible ecological effects (Okundamiya and Ojieabu 2017). Energy diversification, substituting fossil fuels with renewable sources, can eliminate pollutant emissions. This has made renewable power sources a promising field of exploration (Tang et al. 2018; Bukar and Tan 2019; Okundamiya and Omorogiuwa 2016; Ghenai and Bettayeb 2019; Sun and Leto 2020). Also, people are becoming progressively aware of the need to shift towards a greener society with zero-carbon energy harvesting paradigms (IRENA 2020). Consequently, the globe is experiencing a remarkable change in the methods of energy production, transformation, storage, and utilisation in its diverse forms.

Several authors have utilised the tabu search heuristic for optimising hybrid energy systems. In Arefifar et al. (2017), an energy management success index was introduced. A probabilistic-based index expression using the tabu search heuristic was used to minimise the overall operational cost and power losses of hybrid energy systems. The tabu search algorithm was also applied to resolve the energy mix optimisation problem of a microgrid system (Redouane et al. 2018). The aim was to satisfy the energy requirements of an urban centre in Morocco. The results showed a small portion of renewable sources can improve the levelised cost of energy (COE) for different optimum energy mixes.

Puranen et al. (2021) studied the sustainability of a standalone photovoltaic (PV) power system for northern climate conditions. To achieve an optimised off-grid system, a residential disconnected residence in Finland was employed as a case study. The technical viability and optimum configuration of the battery and hydrogen (H_2) storage were determined using data sources from the existing PV installation. The proposed system was analysed using Matlab programming. However, simple models were employed in this study without considering the losses in the storage stages, weather forecasts, consumption estimates, and cost-effectiveness.

Jamshidi and Askarzadeh (2018) jointly considered the co-existence of a PV, diesel generator and fuel cell (FC) system to generate electric power for a remote community. Nevertheless, the presence of operating reserves increases the system's size considerably. A hybrid system integrating PV/FC/battery for the design of a stratospheric airship was proposed by Liao et al. (2018). The goal was to effectively manage the energy while maximising the system's efficiency. Results showed that the system's efficiency and battery life are considerably enhanced. Nonetheless, in the event of battery failure, the capacity of the FC is insufficient to withstand the wind drag of the airship, resulting in increased complexity of the trajectory design. The demand for batteries in microgrids is soaring to an unprecedented level owing to increasing load demand. Hence, hybrid renewable sources (PV, wind turbine, fuel cell) alongside a battery system with a smart battery charge controller aimed at maximising the efficacy and response of batteries implemented using Matlab/Simulink was proposed in Ahmed and Mohamed (2019). Results showed a significant performance gain with the proposed model compared to the conventional charging method. Alvarez-Mendoza et al. (2018) attempted to minimise the impact of frequency instability in hybrid systems by distributed generation. To achieve this goal, a semi-dispatchable energy system designed to stabilise the power output of the hybrid system was proposed. However, the FC nominal total value and the hydrogen flow strongly impact the semi-dispatchability of the hybrid system.

Lagorse et al. (2008) assessed the economic performance of diverse designs comprising PV generators and FC systems but did not include the grid. Ileberi et al. (2016) studied the prospect of distributing green energy to the electrical load of a satellite technology development centre by using the Hybrid Optimisation Model for Electric Renewable (HOMER) software simulation tool. The authors observed that solar and wind energies can be harnessed for use in the study area, but the grid-only system seems very cost-effective with the least net present cost (NPC) and COE. The authors in Kotb et al. (2021) noted the Egyptian Vision 2030, some of which include the augmentation of existing renewables and the drastic increment in the amount of freshwater through the use of desalination plants fed from either brackish or seawater. The HOMER technique was explored to develop an energy-economic-ecological optimisation analysis of nine different configurations. However, sensitivity analysis identifies the enormous growth of reverse-osmosis demand and low-interest rate as a major bottleneck for future investments. The authors identified the intermittency and randomness of renewable energy sources and energy consumption as a major bottleneck mitigating the effective day-to-day operation of residential houses not connected to the electric grid.

There is a gap in the literature on the technical, economic, and environmental analysis and evaluation of a grid-coupled hybrid system for Nigerian cities. This knowledge gap has motivated the present study. This study examines the prospects of using hybrid renewable power resources to augment an existing grid/diesel option for electric power generation. The aim is the analysis and optimisation of the grid/PV/FC/diesel hybrid power system for a reliable and cost-efficient power supply on Nigerian University campuses. The specific objectives are to design a grid-linked PV/FC/diesel hybrid system, simulate and determine the optimal capacity of the hybrid power system for sustainable power supply, and verify the viability of the designed hybrid system. This will facilitate the shift from the usage of the grid/diesel energy harvesting paradigm to a more sustainable power pathway, a crucial transition that would significantly enhance the push toward decarbonising African cities.

6.2 Methodology

The design of the grid-linked photovoltaic/fuel cell microgrid system on the university campus is presented in Fig. 6.1. It consists of a solar photovoltaic system, a fuel cell system integrating an electrolyser with a

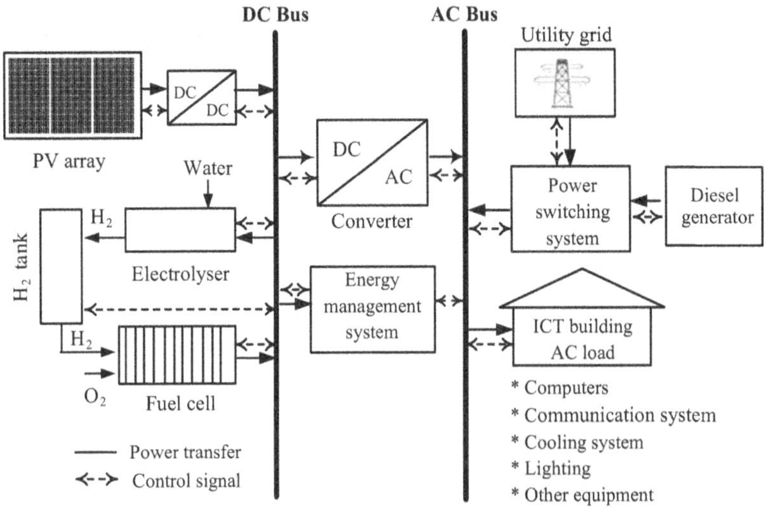

Fig. 6.1 Design of grid-linked PV/FC/diesel hybrid system for ICT load

hydrogen tank, and the DC/AC converters coupled to the grid/diesel power system via bus bars. The power system was designed using HOMER, a computer-based, exhaustive search, dynamic simulation tool for hybrid system sizing with detailed sensitivity assessment, offering the simplicity of operation and use. In addition, it has a good meteorological database.

6.2.1 *Electric Grid Design*

The electrical load of the information and communication technology (ICT) building understudy is linked to the power grid, controlled by the Benin Electricity Distribution Company. The design constraints for developing the erratic grid distribution (Fig. 6.2) for the study area, were deduced as shown in Table 6.1 (Okundamiya 2021b).

6.2.2 *Solar Photovoltaic System*

The output of a photovoltaic system is computed using Eq. (6.1):

$$P_{PV} = f_{PV} Y_{PV} \left(I_T / I_S \right), \tag{6.1}$$

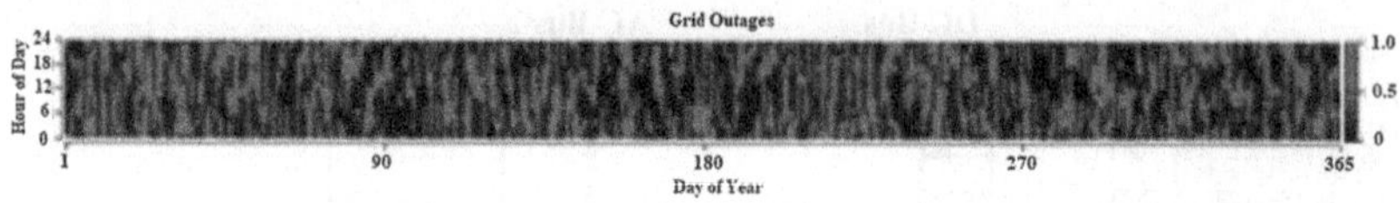

Fig. 6.2 Simulated grid electricity distribution for study site

Table 6.1 Constraints for the erratic grid electricity distribution design

Parameter	*Value*
Mean outage frequency	598
Mean time to repair, MTTR (h)	7.68
Repair time variability (%)	93
Electricity purchase price ($/kWh)	0.1261
Cost of energy sell back to the grid ($)	0
Interconnection fee ($)	1000

Note: $1 ≈ ₦410.35 (Central Bank of Nigeria, accessed September 1, 2021)

f_{PV} is the photovoltaic de-rating fraction, Y_{PV} (kW) is the rated size of the array, I_T (kW/m^2) is the global solar irradiation incident on the surface of the array, and I_S (kW/m^2) is the standard amount of irradiation used to rate the photovoltaic system, typically given as 1 kW/m^2. The de-rating factor accounts for the effects of elevated temperature, wire losses, and dust particles on the performance of the photovoltaic system (Okundamiya 2021a). The techno-economic description of the selected photovoltaic module has been examined in Okundamiya (2021b); the design has an installation cost of $500/kW, the replacement cost of $450/kW, and an operation and maintenance (O&M) cost of $1/kW/y.

6.2.3 *Fuel Cell System*

The electrolyser and the hydrogen tank modules are combined to design an efficient fuel cell system in HOMER (Okundamiya 2021a). An electrolyser technology is essential to generate hydrogen (H_2). The water electrolyser is employed in the separation of water molecules into oxygen (O_2) and H_2 via the application of an electric current. The fundamental constituents of water electrolysers can be disintegrated in three levels: cell, stack, and system (IRENA 2020), as illustrated in Fig. 6.3. The main theory of the cell comprises two electrodes set apart in an electrolyte, which transports chemical charges produced from one electrode to another. The cell, being the hub, where the electrochemical process takes place comprises two electrodes

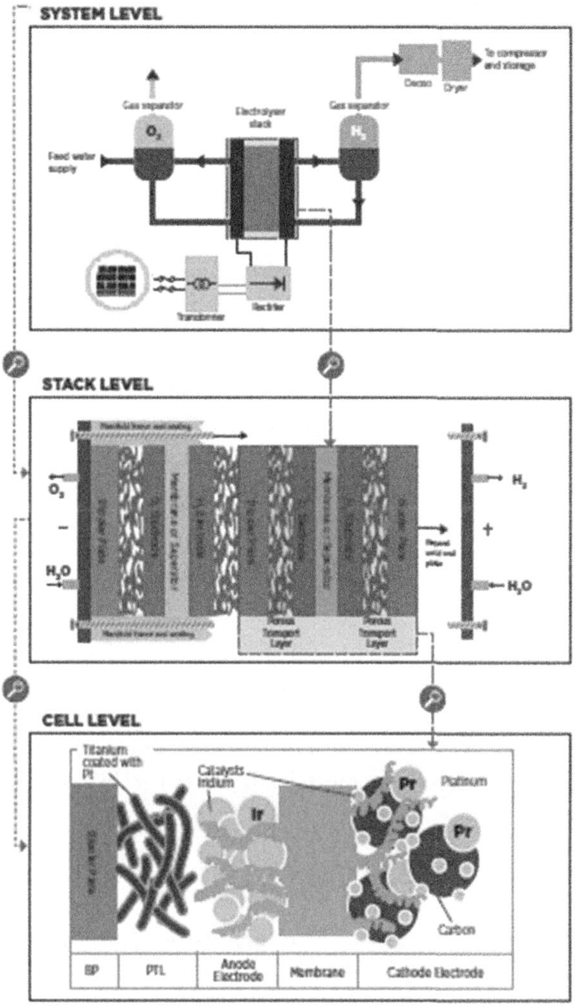

Fig. 6.3 Fundamental constituents of water electrolysers fragmented at different levels (IRENA 2020)

(anode and cathode) that are either submerged in a liquid electrolyte or adjacent to a solid electrolyte membrane, bipolar plates (which dispense the flow offer and mechanical support), and two permeable transport layers that facilitate the reactants transfer as well as products removal. The stack, in contrast, has a wide-ranging scope comprising several cells linked in series, spacers (i.e., insulation between opposite electrodes), mechanical support, seals, as well as endplates, which prevent leaks, take in fluids, etc. Finally, the system goes beyond the stack to embrace mechanisms for cooling, hydrogen processing, conversion of electrical power input, water supply treatment, and gas output.

The excess electric power production of the photovoltaic system is used to energise the electrolyser to generate H_2, for storage in the H_2 tank, by electrolysis of water as follows:

$$H_2O \rightarrow \frac{1}{2}O_2 + H_2. \tag{6.2}$$

The fuel cell system produces electricity via the stored H_2, and O_2, while releasing heat and water (as waste products) as follows (Bukar and Tan 2019):

$$\frac{1}{2}\mathrm{O_2} + \mathrm{H_2} \rightarrow \mathrm{H_2O} + \mathrm{Electricity} + \mathrm{Heat} \tag{6.3}$$

The H_2 fuel can be utilised to satisfy the energy demand either during a shortfall in the PV power output or when there is a high-load demand. The correlation between the electrical power output of the fuel cell with H_2 consumption as well as the efficiency is described in Okundamiya (2021a). The cost details of modules used in the FC system design are shown in Table 6.2.

6.2.4 *Diesel Generator*

The fuel combustion (*L*) of a diesel generator (DG), at time *t*, can be expressed using Eq. (6.4).

$$\mathrm{Fg}(t) = \mathrm{x.Pr} + \mathrm{y.Po}. \tag{6.4}$$

P_r is the rated power (kW), P_o is the output power at time *t* (kW), *x* is the fuel curve intercept coefficient (L/h), and *y* is the fuel curve slope (L/h/kW_o) (Okundamiya 2021b). The characteristic curves (fuel and efficiency)

Table 6.2 Cost details of modules used in the FC system design (Okundamiya 2021b; IEA 2019)

Component	*Parameter*	*Specification*
Fuel cell	Model	Generic
	Capital cost ($/kW)	2000
	Replacement cost ($/kW)	1500
	O&M cost ($/h/kW)	0.015
	Lifecycle (h)	50,000
	Fuel type	H_2
	Search space (kW)	0–150
Electrolyser	Model	Generic
	Capital cost ($/kW)	500
	Replacement cost ($/kW)	250
	O&M cost ($/kW)	5
	Efficiency (%)	75
	Lifespan (y)	20
	Search space (kW)	0–150
Hydrogen tank	Model	Generic
	Capital cost ($/kg)	100
	Replacement cost ($/kg)	80
	O&M cost ($/kg)	1
	Lifespan (y)	25
	Search space (kg)	0–150

of the DG used for the simulation are illustrated in Fig. 6.4. A minimum load ratio of 25% is anticipated for the generator (Okundamiya 2021a).

The lifespan of a DG can be estimated with Eq. (6.5), where, $L_{g,h}$ is the lifespan (h) and N_g is the annual operation time (h/y) (Farret and Simões 2006).

$$L_g = L_{g,h} / N_g. \tag{6.5}$$

The economic description of the DG used in the design is shown in Table 6.3.

6.2.5 *Study Area, Data Collection, and Analysis*

A description of the meteorology of the study area (lat. 6°44.0′N, long. 6°5.1′E) can be found in Okundamiya et al. (2014). The case study meteorological datasets employed for simulation were retrieved from the National Aeronautics and Space Administration (NASA) (NASA (National Aeronautics and Space Administration), Prediction of Worldwide Energy

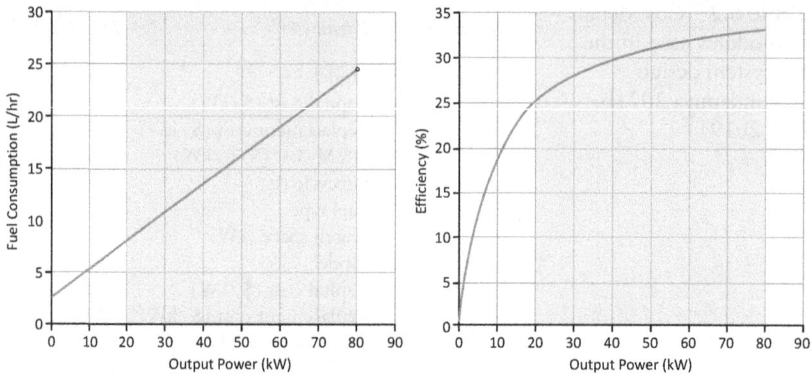

Fig. 6.4 Fuel and efficiency curves of DG used in this study

Table 6.3 Economic description of DG utilised in the design (Okundamiya et al. 2022)

Parameter	*Specification*
Capital cost ($/kW)	370
Replacement cost ($/kW)	300
O&M cost ($/h/kW)	0.15
Fuel cost ($/L)	0.60
Capacity (kVA/kW)	100/80
Lifespan (h)	15,000

Resource (POWER) database). The data comprise monthly averages for global horizontal radiation from July 1984 to June 2005. The data collected was checked for consistency to ensure data accuracy (Table 6.4).

The ICT building's electric load at Ambrose Alli University (AAU) is in use predictably every day of the week from 8.00 to 17.00 hours. The daily electric load utilisation encompasses computers, communications setup (servers, switches, routers, Wi-Fi radios, etc.), lighting and cooling systems, etc. (Fig. 6.5).

The simulated load profiles (Fig. 6.6), with a peak electrical power of 71.20 kW and an annual daily average electric load of 403.43 kWh/d, were developed from the examined data using the HOMER tool (Version 3.14.4) by taking different time adaptability of electric load; hourly variation (15%) and daily variation (10%). Table 6.5 describes the power converter/controller modules costs used in the hybrid power system design (Fig. 6.1). An inflation rate (13.5%), nominal discount rate (8%), and life span (25 y) are assumed for this project.

Table 6.4 22-Year monthly averages for global horizontal radiation for study site (NASA (National Aeronautics and Space Administration), Prediction of Worldwide Energy Resource (POWER) database)

Months	*Daily radiation (kWh/m²/day)*	*Clearness index*
January	5.56	0.598
February	5.60	0.568
March	5.39	0.521
April	5.16	0.494
May	4.86	0.476
June	4.44	0.445
July	3.94	0.392
August	3.82	0.372
September	4.04	0.391
October	4.50	0.452
November	5.06	0.539
December	5.32	0.586

Fig. 6.5 Analysis of the typical daily electric load profile of the ICT infrastructure under study

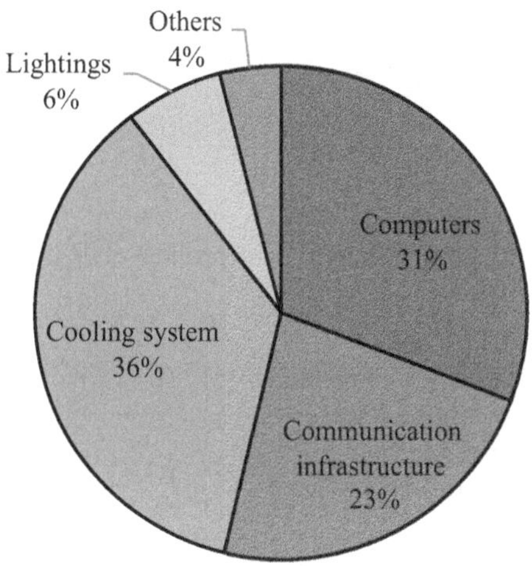

6.3 Results and Discussion

The simulation results are compared with the baseline (conventional grid/DG) system. As observed (Table 6.6), the optimum size design embraces a PV array (340 kW), and FC system (22 kW) integrating an electrolyser (21 kW) with an H_2 storage tank (28 kg) and power electronic converter (80 kW) can sustainably augment the conventional grid-diesel generator system for 0.06235$/kWh. The contribution of the individual subsystems

a

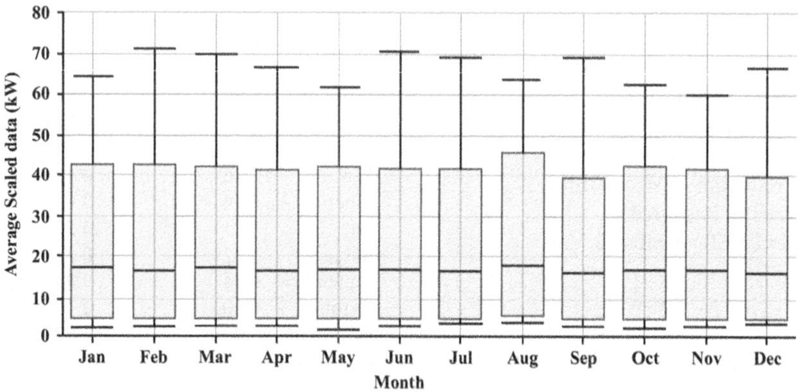

b

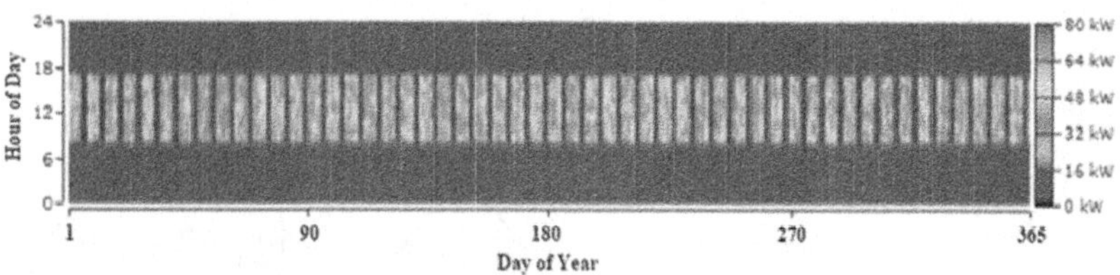

Fig. 6.6 Electric load profiles of ICT infrastructure under study (**a**) seasonal (**b**) yearly

Table 6.5 Economic description of power converter/controller used in system design (Okundamiya et al. 2022)

Component	*Parameter*	*Specification*
Converter	Capital cost ($/kW)	100
	O&M cost ($/kW)	0.50
	Efficiency (%)	95
	Life span (y)	25
	Search space (kW)	0–120
Controller	Capital cost ($)	100
	O&M cost ($/y)	1
	Life span (y)	25

to the overall annual electricity generation of the proposed hybrid system design is presented in Table 6.7. The PV subsystem generates the highest yearly electrical output of 512,202 kWh at a capacity factor of 17.2%. This subsystem operates for 4416 hours with a daily average power output of

Table 6.6 Comparison of architecture of proposed design (grid/PV/FC/DG) and baseline (grid/DG) hybrid system

Component (size)	*Baseline system*	*Proposed system*
Grid (kW)	6000 (7.5 MVA)	6000
PV (kW)	–	340
FC (kW)	–	22
DG (kW)	80 (100 kVA)	80
Electrolyser (kW)	–	21
Power converter (kW)	–	80
H_2 tank (kg)	–	28

Table 6.7 Contribution of the individual subsystems to the overall annual electricity generation of the proposed hybrid system design (grid/PV/FC/DG)

Subsystem	*Annual electricity production*	
	kWh/y	*%*
PV	512,202	93.08
FC	24,966	4.54
DG	2327	0.42
Grid	10,791	1.96
Total	550,286	100.00

1403 kW. The FC, which contributes only 4.54% of the total electrical output, operates for 3186 hours per annum with H_2 consumption of 1104 kg to generate a mean electrical power output (7.84 kW). The electrolyser utilises 59,467 kWh/y (electricity) to generate H_2 (1131 kg/y) at a specific consumption rate of 52.6 kWh/kg.

A comparison of the techno-economic merits of the proposed and baseline hybrid power systems is shown in Table 6.8. As noticed, the integration of the PV/FC hybrid power source with H_2 storage to the traditional grid/DG system enhances the renewable portion by 91.1%. Consequently, the overall annual grid electricity purchases reduce by 84% (from 62,274 kWh/y to 10,791 kWh/y), whereas the duration of diesel generator operation reduces by over 97% (from 4748 h/y to 108 h/y).

A significant change (40%) is noticed in the electrical power production and utilisation of both power system pathways. The 40% discrepancy in electricity consumption (i.e., 59,467 kWh/y) is attributable to the losses from the conversion of direct current to alternating current power being proposed in the present design. It is imperative to mention that a power system that incorporates renewable sources such as solar needs more operating reserves to protect against arbitrary decreases in the renewable energy supply. Hence, compared to the baseline system, the proposed

Table 6.8 Comparison of techno-economic merits of proposed and baseline systems

Techno-economy	*Baseline system*	*Proposed system*
Renewable fraction (%)	0	91.1
Grid energy purchases (kWh/y)	67,274	10,791
DG Fuel (L/y)	47,144	921
DG Operation (h/y)	4748	108
Annual energy production (kWh/y)	194,047	550,286
Annual energy utilisation (kWh/y)	147,252	206,719
Capacity Shortage (%)	0	0
Unmet load (%)	0	0
Excess electricity (%)	24.1	61.1
Initial capital ($)	30,700	266,000
Net Present Cost ($)	5,116,513	466,434
Operating cost ($/y)	100,109	3945
COE ($/kWh)	0.684	0.06235

hybrid system generated higher excess electricity (61.1%) to sustain the electric load demand of the ICT infrastructure under study.

The economic viability of a power system can be assessed using two key indices: the COE and NPC of the power system (Okundamiya 2021b). The system design, which satisfies the electricity demand subject to predefined limits (or constraints) with the lowest COE and NPC is considered the most viable design approach. As shown in Table 6.8, the proposed hybrid power system with H_2 storage reduced the NPC of the baseline system by over 90% (from $5,116,513 to $466,434) with a resultant saving in the energy cost (from $0.6840/kWh to $0.06235/kWh). Although the proposed system requires a somewhat high initial capital, the exceptionally low annual operating cost indicates an enhanced cost saving compared to the baseline scenario. A comparative analysis of the economics, as well as the cumulative cash flow of the proposed and baseline energy harvesting systems over the project lifetime, are presented in Table 6.9 and Fig. 6.7 respectively.

The present worth suggests that an overall cost reduction of $4,650,079 can be salvaged by installing the proposed hybrid energy system (corresponding to yearly payback of $91,532) as an alternative to the baseline for energy harvesting of the infrastructure under study. An ROI index value (36.6%) showed the economic rewards compared more positively to the baseline. The IRR index measured the time worth in assessing the fund of a project (Okundamiya 2021a). The IRR index value (39.5%)

Table 6.9 Comparison analysis of economics of proposed and baseline systems

Parameter	*Annual worth (\$/y)*	*Present worth (\$)*	*Internal rate of return, IRR (%)*	*Return on investment, ROI (%)*	*Simple payback (y)*	*Discounted payback (y)*
Value	91,532	4,650,079	39.5	36.6	2.64	2.42

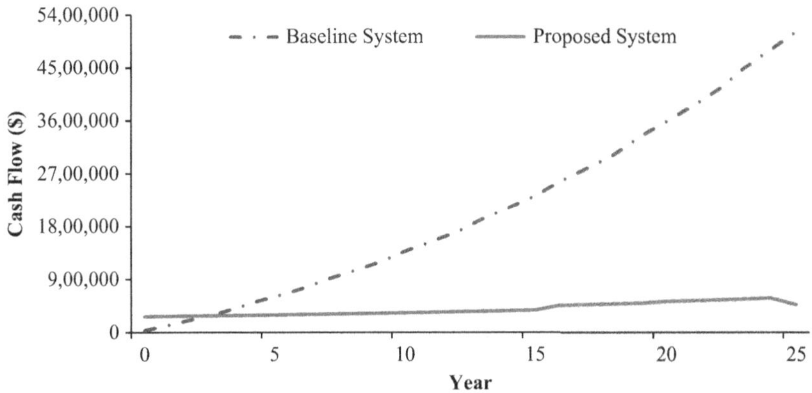

Fig. 6.7 Cumulative cash flow of the baseline and proposed systems compared over project lifespan

shows a viable system with a payback period of fewer than 3 years. Table 6.10 presents the emission characteristics of the developed and baseline systems.

The proposed system reduces the annual pollutant emission from 92.71% (SO_2) to 98.05% (PM) with an average of 96.29%. Specifically, 94.44% CO_2 reduction can be achieved by employing the proposed hybrid system. This significant cutback in carbon emissions of the ICT infrastructure can help in decarbonising the campus environs, thereby making the immediate surroundings more eco-friendly and cleaner.

6.4 Conclusion

This chapter assessed the viability of augmenting a grid/diesel power with a hybrid PV/FC system including an electrolyser and H_2 storage tank for electricity harvesting on a university campus. The objective was to find out

Table 6.10 Pollutant emission characteristics of proposed and baseline systems

Pollutants	*Carbon dioxide (CO_2)*	*Carbon monoxide (CO)*	*Unburned hydrocarbon*	*Particulate matter (PM)*	*Sulphur dioxide (SO_2)*	*Nitrogen oxides (N_2O)*
Baseline system emission (kg/y)	165,933	770	33.9	4.62	487	814
Proposed system emission (kg/y)	9230	15.3	0.663	0.0902	35.5	28.6
Emission reduction (%)	94.44	98.01	98.04	98.05	92.71	96.49

the economic, technical, and ecological gains of the integration since renewable power resources are location-dependent. The energy-balance methodology was employed for the power system design. Case study simulations were analysed for the electric load of an ICT infrastructure at a university campus (lat. 6°44.0′N, long. 6°5.1′E) using 22-year period datasets gathered from NASA. It was observed that the integration of a PV array (340 kW) and FC system (22 kW) incorporating an electrolyser (21 kW) with an H_2 storage tank (28 kg) and power electronic converter (80 kW) can sustainably boost the conventional grid-diesel generator system for 0.06235$/kWh. Consequently, the overall annual grid electricity purchases reduce by 84% whereas the renewable fraction increased by 91.1% and the duration of diesel generator operation reduce by over 97%. The implementation can help to resolve the energy setbacks hindering sustainable development in Nigerian universities. Besides, the hybrid system could alleviate the emission of CO_2 from other power sources, making the environment cleaner.

The deployment of this application to Africa's campus cities, and indeed the main commercial hub cities, would mean that the cost savings and carbon abatement benefits can be multiplied several times. The evidence from this modelling of the potential applications of the hybrid system can be used in the engagement and negotiations with other stakeholders and policymakers. This will help to prioritise specific enablers like having relevant policies in place or designing/reconfiguring the grid to which the micro-grid will be linked. In this respect, the hybrid system will help accelerate the decarbonisation of African cities while offering resource-rich countries an orderly or phased transition from fossil fuels given that some fossil fuels would still be required in the hybrid system.

List of Abbreviations

AAU	Ambrose Alli University
AC	Alternating current
CO	Carbon monoxide
CO_2	Carbon dioxide
COE	Cost of energy
DC	Direct current
DG	Diesel generator
FC	Fuel cell
H_2	Hydrogen
HOMER	Hybrid optimisation model for electric renewables
ICT	Information and Communication Technology
IRR	Internal rate of return
MTTR	Mean time to repair
NASA	National Aeronautics and Space Administration
N_2O	Nitrogen oxide
NPC	Net present cost
O&M	Operation and maintenance
O_2	Oxygen
PM	Particulate matter
PV	Photovoltaic
ROI	Return on investment
SO_2	Sulphur dioxide
SSA	Sub-Saharan Africa

References

M. S. Okundamiya, "Size optimization of a hybrid photovoltaic/fuel cell grid connected power system including hydrogen storage," International Journal of Hydrogen Energy, vol. 46, no. 59, pp. 30539-30546, 2021a

M. S. Okundamiya. "Integration of photovoltaic and hydrogen fuel cell system for sustainable energy harvesting of a university ICT infrastructure with an irregular electric grid", Energy Conversion and Management, vol. 250, no. 114928, 2021b, https://doi.org/10.1016/j.enconman.2021.114928

Energy Access Outlook 2017: From Poverty to Prosperity, World Energy Outlook Special Report. Available from https://www.gogla.org/sites/default/files/resource_docs/weo2017specialreport_energyaccessoutlook.pdf (accessed on 10 January 2019)

IRENA, "Global Energy Transformation: A roadmap to 2050," International Renewable Energy Agency, Abu Dhabi, 2018.

M. S. Okundamiya, C. E. Ojieabu, "Optimum design, simulation and performance analysis of a micro-power system for electricity supply to remote sites," Journal of Communications Technology, Electronics and Computer Science, vol. 12, pp. 6–12, 2017.

S. Tang, H. Honga, J. Sunc, W. Qu, "Efficient path of distributed solar energy system synergetically combining photovoltaics with solar-syngas fuel cell," Energy Conversion Management, vol. 173, pp. 704–714, 2018.

A. Bukar, C. W. Tan, "A review on stand-alone photovoltaic-wind energy system with fuel cell: System optimization and energy management strategy," Journal of Cleaner Production, vol. 221, pp. 73-88, 2019.

M. S. Okundamiya, O. Omorogiuwa, "Analysis of an isolated micro-grid for Nigerian terrain," 59th Int Midwest Symposium on Circuits & System Abu Dhabi, UAE, 2016.

C. Ghenai, M. Bettayeb, "Modelling and performance analysis of a stand-alone hybrid solar PV/fuel cell/diesel generator power system for university building," Energy, vol. 171, pp. 180-189, 2019.

C. Sun, S. Leto, "A novel joint bidding technique for fuel cell wind turbine photovoltaic storage unit and demand response considering prediction models analysis effect's," International Journal of Hydrogen Energy, vol. 45, pp. 6823-6837, 2020

IRENA, "Green Hydrogen Cost Reduction: Scaling up Electrolysers to Meet the 1.5^{0}C Climate Goal," International Renewable Energy Agency, Abu Dhabi, 2020.

S. A. Arefifar, M. Ordonez, Y. A.-R. I. Mohamed, "Energy management in multi-micro grid systems development and assessment," IEEE Trans. Power Syst., vol. 32, no. 2, pp. 910–922, 2017.

Abdelbari Redouane, Imad El Harraki, and Mohamed Ouzineb, "Energy mix capacity factor optimization using Tabu search algorithm," 1st Int'l Congress on Solar Energy Research, Technology and Applications, in AIP Conference Proceedings 2056, 020001 (2018); https://doi.org/10.1063/1.5084974, December 2018.

P. Puranen, A. Kosonen, and J. Ahola, "Technical feasibility evaluation of a solar PV based off-grid domestic energy system with battery and hydrogen energy storage in northern climates," Sol. Energy, vol. 213, 2021, https://doi.org/10.1016/j.solener.2020.10.089.

M. Jamshidi and A. Askarzadeh, "Techno-economic analysis and size optimization of an off-grid hybrid photovoltaic, fuel cell and diesel generator system,"

Sustainable Cities and Society, https://doi.org/10.1016/j.scs.2018.10.021, 2018.

J. Liao, Y. Jiang, J. Li, Y. Liao, H. Du, W. Zhu and L. Zhang, "An improved energy management strategy of hybrid photovoltaic/battery/fuel cell system for stratospheric airship," Acta Astronautica, vol. 152, pp. 727-739, 2018.

F. B. Ahmed and M. I. Mohamed, "Battery charge management for hybrid PV/wind/fuel cell with storage battery," Energy Procedia, vol. 162, pp. 107-116, 2019.

F. Alvarez-Mendoza, C. Angeles-Camacho, P. Bacher, and H. Madsen, "Semi-dispatchable generation with wind-photovoltaic-fuel cell hybrid system to mitigate frequency disturbance," Electric Power Systems Research, vol. 165. pp. 60-67, 2018.

J. Lagorse, M. G. Simoes, A. Miraoui, P. Costerg, "Energy cost analysis of a solar-hydrogen hybrid energy system for stand-alone applications," International Journal of Hydrogen Energy, vol. 3, pp. 2871–2879, 2008.

R. G. Ileberi, O. H. Adikankwu, A. E. Timi, I. O. Adedekan, "Grid integration of renewable technology: A techno-economic assessment," American Journal of Mechanical Engineering, vol. 4, no. 5, pp. 182–190, 2016.

K. M. Kotb et al., "A fuzzy decision-making model for optimal design of solar, wind, diesel-based RO desalination integrating flow-battery and pumped-hydro storage: Case study in Baltim, Egypt," Energy Convers. Manag., vol. 235, 2021, https://doi.org/10.1016/j.enconman.2021.113962.

IEA. The Future of Hydrogen. Tech. rep. Paris, 2019. URL: https://www.iea.org/reports/the-future-of-hydrogen.

F. A. Farret, M. G. Simões, "Micropower system modelling with Homer," in Integration of Alternative Sources of Energy, IEEE, 2006, pp. 379-418, https://doi.org/10.1002/0471755621.ch15.

M. S. Okundamiya, S. T. Wara, H. I. Obakhena, "Optimization and techno-economic analysis of a mixed power system for sustainable operation of cellular sites in 5G era," International Journal of Hydrogen Energy, vol. 47, no. 39, pp. 17351-17366, 2022.

M. S. Okundamiya, V. O. A. Akpaida, B. E. Omatahunde, "Optimization of a hybrid energy system for reliable operation of automated teller machines," Journal of Emerging Trends Engineering & Applied Sciences, vol. 5, no. 8, pp. 153-158, 2014.

NASA (National Aeronautics and Space Administration), Prediction of Worldwide Energy Resource (POWER) database, Monthly averages for global horizontal radiation over 22-year period (from July 1983 – June 2005), available online http://eosweb.larc.nasa.gov/; (accessed on January 31, 2021).

PART II

Governance and Policy Approaches for Decarbonising African Cities

CHAPTER 7

Powering Action Towards Energising African Cities Sustainably: Perspectives from Kenya

Laura Muniafu, Peter Ombewa, and Nzembi Mutiso

Abstract The drive for cleaner energy sources presents opportunities and challenges for developing countries with hydrocarbon resources. Although reducing carbon emissions that arise from fossil fuel consumption will be a plus, resource-rich African countries are bound to lose out on resource rent, which will impact their development agenda. Energy-sourcing is now an intricate art of balancing between "*variable-clean*" energy sources and "*proven-dirty*" fossil fuels, just like cities the world over, which consume 80% of energy production and produce 75% of all the emissions. African cities must ensure that population and economic growth is energised sustainably to address challenges of energy poverty, while meeting

L. Muniafu (✉)
Strathmore Extractive Industry Centre, Strathmore University, Nairobi, Kenya
e-mail: lmuniafu@strathmore.edu

P. Ombewa
Eaton Corporation, Nairobi, Kenya

N. Mutiso
Extractives Baraza, Strathmore University, Nairobi, Kenya

S. I. Azubuike et al. (eds.), *Decarbonisation Pathways for African Cities*, Palgrave Studies in Climate Resilient Societies,
https://doi.org/10.1007/978-3-031-14006-8_7

the increasing energy demands of rapidly growing economies. This chapter focuses on decarbonisation in Kenya and considers steps key stakeholders are taking towards decarbonisation.

Keywords Decarbonisation • Climate change • Kenya • Local innovation

7.1 Introduction

The world population is on the rise, putting pressure on the available resources. An interesting trend in this growth is the increasing population within urban areas owing to rising rural-urban migration and reclassification of formerly rural settlements into urban centers. For instance, according to the United Nation's World Population Prospects (UN 2014), 54% of the global population resides in urban settlements, up from 30% in 1950. By 2050 this figure is estimated to reach 66%.

The Population Research Bureau (PRB) publishes annually the *World Population Datasheet*, which captures demographic information for more than 200 countries and territories (PRB 2021). According to the data published for 2021, the world's population stood at about 7.8 Billion people in mid-2020 and is expected to rise to 8.9 Billion and 9.9 Billion in 2035 and 2050 (PRB 2021), growing at an average rate- of natural increase of 1.2%. According to the same data, Africa's population is expected to reach 2.6 Billion in 2050, up from the current 1.9 Billion as of mid-2020. Africa's rate-of-natural increase stands at 2.6%, more than double the global average. The PRB data also indicates that 56% of the world's population and 43% of the African population live in urban areas (PRB 2021).

The rate of urbanisation in Africa is relatively high compared to other major regions, competing only with Asian countries. In 2020, 43% of the African population was classified as urban, up from 27% in 1950, with projections of up to 60% by 2050. A ranking by (Teye 2018) shows that the top-ten largest African cities by population include: Lagos, Nigeria (21 million people), Cairo, Egypt (20.4 million), Kinshasa, DRC (13.3 million), Luanda, Angola (6.5 million), Nairobi, Kenya (6.5 million), Abidjan, Cote d' Ivoire (4.8 million), Alexandria, Egypt (4.7 million), Johannesburg,

South Africa (4.4 million), Dar es Salaam, Tanzania (4.4 million), and Casablanca, Morocco (4.3 million).

Urbanisation is associated with increased incomes and purchasing powers, which leads to an increase in the demand for energy and increased greenhouse gas (GHG) emissions. According to data cited by the Intergovernmental Panel on Climate Change (IPCC), in 2006, urban areas accounted for 67–76% of energy use and 71–76% of energy-related CO_2 emissions (IPCC 2014).

World energy production is considered to be still highly dependent on the burning of fossil fuels (Ritchie and Roser 2020). 33% of the global primary energy is evidently derived from oil, 27% from coal, and 24.2% from gas, all of which are fossil fuels associated with high levels of GHG emissions. Only 16% of the energy consumed in the world comes from renewable sources. Africa's case is not any different, with a total consumption of about 5519 TWh (2019), 2300 TWh was derived from oil, 1501 TWh from gas, and 1242 TWh from coal (Ritchie and Roser 2020).

The burning of fossil fuels and industrial processes were among the leading sources of global CO_2 emission between 1970 and 2010, contributing about 78% of total GHG emissions (IPCC 2014). Energy, electricity, and heating sectors contributed about 34% of the total global direct GHG emissions in 2010 and 1.4% of the indirect emissions for the same period (IPCC 2014).

The growth of African cities is bound to continue driven by increasing rural to urban migration, reclassification of settlement, and improving living standards. As a result, the growing cities will see an increase in energy demands. Therefore, governments across the continent will need to invest in energy infrastructure. Currently, (coal, oil, and gas) are the major primary sources of energy for many African countries. Left unchecked, this heavy dependence on fossil fuel in Africa poses a huge threat to combating climate change and minimising its impacts. Therefore, the growing African cities must come up with ways to energise sustainably.

This chapter focuses on the Kenyan case. It will present the country's circumstances regarding climate change impacts and analyse some of the efforts in terms of policies, laws, institutions, and international collaboration put in place by Kenya to ensure its cities are powered sustainably.

7.1.1 *Kenya's GHG Context (or Kenya's Experience with GHG)*

The population of Kenya is 47.9 million people as of 2019. The population has increased by 26% over a period of ten years from 37.7 Million in 2009. About 14.7 Million people lived in urban areas as of 2019 representing 31% of the population. The major cities include Nairobi and its environs (5.5 Million), Mombasa (1.2 Million), Nakuru (0.6 Million), Eldoret (0.5 Million), and Kisumu (0.4 Million) (KNBS 2019).

The PRB estimates Kenya's rate of natural increase in population at 2.3%. With this rate of growth, Kenya's population is expected to hit 71.9 Million and 89.7 Million by mid-2035 and 2050, respectively (Bureau, Rate of Natural Increase, 2021). Kenya's gross domestic produce (GDP) stood at $82 Billion (2017). Experts have estimated that adverse impacts of climate change have seen the country lose between 3–5% of the GDP over the past decade (UNFCCC 2020). In 2016, Kenya's energy consumption stood at 6122 thousand tonnes of oil equivalent, of which 5.6% was derived from coal and coke, 82.4% from liquid petroleum fuels, and 12% from hydro and geothermal power generation (KNBS 2016).

An in-depth analysis of this data reveals that although Kenya relies heavily on liquid petroleum fuels, only a small portion goes directly to the production of power. For instance, in 2016, 73.7% of the liquid fuels consumed locally were sold at retail petroleum outlets and on-road transport, 11.9% sold as aviation fuels, and 12.2% sold to industrial customers. Only 15.3 of the 5044 thousand tonnes consumed that year went directly into power generation, representing about 0.3% compared to 3.3% in 2012 when Kenya experienced a drought (KNBS 2016).

In 2015, Kenya's greenhouse gas emissions increased from 56.8 $MtCO_2e$ in 1995 to 93.7 $MtCO_2e$, representing an increase of about 65% over a period of ten years. It is projected that this value will rise to 143 $MtCO_2eq$ by 2030 as the country endeavours to attain economic goals spelled in its Vision 2030 plan. (UNFCCC 2020). The leading sources of GHG emissions in Kenya by sector as of 2015 include;

- Agriculture, which accounts for 40% of Kenya's emissions. Agricultural emissions come from livestock fermentation, manure, and application of fertiliser)
- Land use and land-use change and forestry (LULUCF) emissions which accounts for 38% owing to widespread deforestation

- Energy, including transport accounting for 18%. It is projected that energy and transport-related emissions will lead to GHG emissions in Kenya by 2030.
- Industrial processes and product use (PPU), accounting for 3%
- Waste management accounting for 1%

Even though Kenya's contribution to global GHG emission might be considered low (0.1% of global emissions), the country has not been spared from climate change's harsh effects. While submitting the Updated NDC, 2020, Kenya listed the following as already affecting the nation's socio-economic wellbeing:

- Increase in surface temperatures—According to evidence presented in IPCC 2014 report, Kenya's surface temperatures have risen by 0.5–2 °C over the past 100 years, and since 1950 the country has experienced an altered magnitude and frequency of extreme climate events owing to climate change. The surface temperature is expected to rise by another 1–1.5 °C by 2030 if left unchecked
- There is a decrease in frequency of cold days and nights and hot days and nights, including heatwaves.
- Interruptions in the regional rainfall projection include a significant increase in October-December short rains in many counties of Kenya and drops in March-May during long rains. Northern Kenya has been experiencing substantial rainfall deficits leading to droughts, while the southern region has recorded a slight increase in rainfall. Existing studies predict a decline of up to 100 mm in precipitation levels in the long rain season.
- There have been increasing droughts and floods in different parts of Kenya, causing losses of lives and livelihoods. For instance, in 2018, floods experienced in the country resulted in more than 230,000 people being displaced, closure of over 700 schools, loss of close to 8500 hectares of crops, and destruction of billions worth of road infrastructure.
- Droughts experienced between 2014–18 affected 23 of Kenya's 47 counties and left 3.4 million people food insecure.
- The rise in sea levels affecting coastal towns and communities has also been reported, with an impacted population estimated at 86,000 people a year. In addition, rising sea levels result in coastal erosion

and loss of wetlands, affecting Kenya's tourism sector, one of the country's foreign exchange earners.

- Rising sea temperatures off the Kenyan coast have resulted in massive coral bleaching and mortality on coral reefs system over the last two decades affecting the abundance and composition of the fish species, thus negatively affecting coastal fisheries.
- Increasing temperatures have resulted in the shrinkage of 17% in Mount Kenya's glaciers, threatening one of Kenya's largest river basins, the river Tana, putting to risk the country's more than 60% of hydropower generation capacity.
- Kenya's hydropower generation has also been affected by the irregular rainfall leading to lower than normal water levels on the dams resulting in lower electricity generation. This forces the country to rely heavily on the expensive and CO_2-intensive diesel-thermal power plants for electricity.

Already, Kenya has recorded between 3 and 5% loss in annual GDP caused by adverse climate change. As a result, the country's leadership has embraced the global agenda to combat climate change and adapt to its impacts.

7.1.2 Kenya's Commitment to the Paris Agreement

With full acknowledgement that climate change poses a threat to the country's economic prosperity, Kenya joined 196 other Parties in the 21st Conference of Parties (COP-21) to the United Nations Framework Convention on Climate Change (UNFCCC) held in December 2015 in Paris, France. Kenya ratified the Paris Agreement (PA) a year later in December 2016.

Under Article 4(2) of the PA, each Party is required to prepare, communicate to the UNFCCC, and maintain successive NDCs it intends to achieve (Nationally Determined Contributions (NDCs) | UNFCCC). UNFCCC maintains an interim registry of all the NDCs submitted on its website (UNFCCC 2021a). According to this database, Kenya submitted her first NDC in 2016 under the banner of Kenya National Adaptation Plan (NAP): 2015–2030. Additionally, in line with the UNFCCC requirement that the NDC targets be updated every five years to reflect a more stringent target to reduce GHG emissions, Kenya submitted the first update under the banner Kenya Updated NDC in 2020 (UNFCCC 2020).

The implementation of this NDC was fully conditional to international support in raising the budget required to achieve the targets. Some of the measures proposed to achieve the targets of this NDC included (Kenya First NDC, 2021):

- increasing the contribution of renewable energy sources into Kenya's energy mix through geothermal, solar, and wind energy production;
- making significant progress towards the achievement of growing trees in at least 10% of Kenya's landscape;
- adopting clean energy technologies and reducing the country's over-reliance on wood fuels;
- adopting low carbon and more efficient transportation systems;
- promoting climate-smart agriculture (CSA) in line with the country's National CSA, and;
- encouraging sustainable waste management.

Kenya submitted an updated NDC to the UNFCCC on 28th December 2020, which builds upon the proposals outlined in the first NDC and is modelled as per the guidelines in "*Enhancing NDCs: A Guide to Strengthening National Climate Plans by 2020*" issued by the World Resource Institute and UNDP. The updates also consider proposals captured in Kenya's second National Climate Change Action Plan (NCCAP 2018–2022) and the Third National Inventory Report (NIR3).

Key features of the updated NDC include increased targets for reduction in GHG emissions from 30% to 32% by 2030 relative to the BAU scenario of 143 MtCO_2e. The updated NDC outlines both **mitigation** and **adaptation** measures required to meet the 32% reduction in GHG emissions by 2030.

The **mitigation plan** is expected to cost about USD17.725 Billion, with 21% being sourced locally. Under the measures, the document proposes to implement through the National Climate Change Action Plans (NCCAPs) the following priority activities (UNFCCC 2020):

- achieving 10% of tree cover in Kenya
- making efforts towards achieving land degradation neutrality
- scaling up Nature-Based Solutions (NBS) for mitigation
- reducing the country's overreliance on wood fuels by adopting clean, efficient, and sustainable energy technologies
- promoting low-carbon and efficient transportation systems

- adopting low carbon and more efficient transportation systems.
- promoting climate-smart agriculture (CSA) in line with the country's National CSA framework and with emphasis on efficient livestock management systems
- encouraging sustainable waste management systems.
- harnessing the mitigation benefits of sustainable blue economy and coastal carbon payment for ecosystem services (PES)

According to the updated NDC, the goal of the **adaptation plan** is to ensure a climate-resilient society in Kenya. The adaptation measures will also be linked to Kenya's Vision 2030, NCCAP II (2018–2022), and the NAP (2015–2030). The measures outlined for adaptation are expected to cost the economy some USD43.927 Billion, 10% of which will be sourced locally.

The specific adaptation actions considered in the updated NDC include:

- enhancing the adaptive capacity and climate resilience across all sectors of the economy at both national and county government.
- financing locally led climate change actions to improve climate resilience in local communities
- developing and applying comprehensive climate risk management tools to enhance the risk-based approach to climate change adaptation
- addressing residual climate change impacts, loss, and damage in the productive sectors of the economy
- enhance generation, packaging, and widespread uptake and use of climate information in decision making and planning across sectors and counties with robust early warning systems (EWS)
- enhancing uptake of adaptation technologies, especially among women, youth, and other vulnerable groups
- enhancing investment in blue economy
- strengthening tools for adaptation monitoring, evaluation, and learning (MEL) at all levels of government and amongst non-state actors.

7.2 Legislative and Policy Framework for Climate Change Mitigation in Kenya

Kenya has a comprehensive legislative and policy framework governing the energy sector, whilst factoring in climate change mitigation that has been useful in shaping its approach towards achieving sustainable energy. The table below lists the various legislation and commitments in place to achieve sustainable energy.

7.2.1 Current Legal and Regulatory Provisions

Climate Change Act (2016)	This is the first comprehensive legal framework for climate change governance in Kenya. This Act established the National Climate Change Council to govern climate change in the country chaired by the President. Kenya is one of the countries globally that has attempted to regulate climate change directly.
The Energy Act (2019)	This Act has created a regulatory framework to ensure Kenya achieves its goal of universal energy by 2030. This Act provides for the devolution of energy services to the grassroots level. The energy plans which are supposed to be designed under this Act are to be implemented at the county level. County governments will be compelled under this Act to use locally available resources in the design of energy plans. This Act further empowers county governments to build local renewable energy centres (Gicheru 2019).

7.2.2 Policies, Commitments, and Instruments in Relation to Sustainable Energy

Kenya Vision 2030 (2008) and its Medium-Term Plans (MTP)	This blueprint recognised climate change as a risk to the country and could possibly slow down Kenya's development. Climate change actions were also identified in the Second Medium-Term Plan and Third Medium-Term Plan
Kenya's National Climate Change Response Strategy (2010)	This was the first National Policy document on climate change. This has sought the advancement of climate change adaptation and mitigation.
National Climate Change Action Plan (NCCAP)	This seeks to further the development goals in a low-carbon climate-resilient manner. This plan enables actions for the 2018–2022 period

(*continued*)

(continued)

The National Climate Finance Policy	The National Climate Finance Policy promotes the establishment of legal, institutional, and reporting frameworks for access to, and management of climate finance. In addition, the policy aims to further Kenya's national development goals through enhanced mobilisation of climate finance that contributes to low-carbon climate-resilient development goals.
The National Climate Change Framework Policy	The Policy aims to integrate climate change considerations at the National and County levels (Kurdziel et al. 2019)
Ratification of the Paris Agreement (2016)	This was ratified by Kenya in 2016. The Agreement has provided considerable implications for decarbonisation
Nationally Determined Contribution (NDC) (2016)	This is under the Paris Agreement. However, there is a need for international support to achieve this

From a general viewpoint, Kenya has made significant strides in Sustainable Energy development and management, evidenced by the foregoing legal, policy, and institutional frameworks in place at the national level. As a matter of fact, the electricity sector in Kenya is deemed to be strongly aligned towards the energy transition in comparison to other African and global counterparts given that it has one of the lowest carbon-intensive power systems in the region owing to the fact that >70% of its energy mix is from hydro and geothermal sources (Kahlen et al. 2021). Only time will tell how best the country maintains harnessing its renewable energy potential which is poised to be particularly material-intensive.

7.3 Renewable Energy Potential in Kenya

There is potential in power generation from renewable energy sources. The government has expanded renewable energy generation through solar, hydro, wind, and geothermal. The government has further prioritised the development of geothermal and wind energy plants.

Solar

This is an option for rural electrification and decentralised applications. This subsidises stand-alone systems for households and public institutions. Kenya aims to install an additional 500 MW and 300,000 domestic solar systems by 2030 (GET.Invest 2018).

Wind
Currently, there are 35 meteorological stations in Kenya. It is estimated that about 25% of the country will have wind technology. The largest wind farm in Africa (300 MW) is under construction in the Turkana area of North-Western Kenya. The Ngong hills area close to Nairobi also has 5.1 MW installed, and private investors plan to install several MW of capacity. An average of 80–100 small wind turbines (400 W) have been installed to date, often as part of a hybrid PV-wind system with battery storage (GET.Invest 2018).

Hydro
The projected generation costs for hydro sites are excluded from the Least Cost Power Development Plan (LCPDP). Small, mini, and micro hydroelectric systems (with capacities of less than 10 MW) are estimated to generate 3000 MW nationwide. In 1997, Kenya's Electric Power Act allowed independent power producers to supply electricity to the grid, but small decentralised schemes, such as micro hydropower, were not fully addressed. Around 55 river sites suitable for rural electrification have been identified as attractive commercial opportunities. Their maximum mean capacities would range from 50 kW to 700 kW. The country's agricultural activity produces large amounts of agricultural waste. These can be used to produce electricity by implementing biogas and biomass technologies. The 2014 National Energy Policy Draft also sets out biogas expansion targets of 10,000 small and medium-sized digesters by 2030. Biogas is considered a viable energy solution by a number of agricultural producers (GET.Invest 2018).

Geothermal
Kenya is endowed with geothermal resources, mainly in the Rift Valley. Geothermal and wind energy have comparably low electricity production costs, and the potential capacity is 10 GW for geothermal and a minimum of 1 GW for wind. Conservative estimates suggest geothermal potential in the Kenyan Rift at 2000 MW, whereas the total national potential is between 7000 and 10,000 MW. Production started in 1981 when a 15 MW plant was commissioned in Olkaria. KenGen and an independent power producer currently produce a total of 129 MW. Geothermal power has been identified as a cost-effective power option in Kenya's Least Cost

Power Development Plan. Exploration for geothermal energy in the high-potential areas of the Kenyan Rift is ongoing (GET.Invest 2018).

The electricity sector is also leaning towards decarbonisation. The following are some of the findings made from a research conducted on decarbonisation in the electricity sector:

1. The participation of climate policy planning representatives in electricity sector planning processes could be ensured, and vice versa. The appointed climate change focal point in the Ministry of Energy is based at the Renewable Energy Directorate. The Ministry of Energy and the Ministry of Petroleum and Mining should also be included
2. Increased awareness and certainty on the latest and most relevant electricity sector planning documents
3. Increased use of GHG emissions information included in electricity sector planning documents referred to as Least Cost Power Development Plans (LCPDPs): For the first time, the 2017–2037 Least Cost Power Development Plan (LCPDP) included a dedicated section on the implications of the LCPDP scenarios for GHG emissions and climate change mitigation objectives. This information can be directly used in climate change planning processes to align with information on emission baselines and mitigation options.
4. Establishing longer-term targets for the electricity sector: The LCPDP begins with reference to major national objectives that the plan seeks to fulfil, such as the achievement of universal access to electricity (Kurdziel et al. 2019).

The resource modelling and Life-cycle Assessment(LCA) analysis from a research on the evaluation of GHG savings that may be achieved if sustainably produced biomass briquette fuels are used by both domestic and industrial users. Research has generated a series of results from which key lessons may be drawn (Welfle et al. 2020). These lessons are as follows:

1. Kenya has large potential bioenergy opportunities through utilising the wastes and residues from ongoing Kenyan agricultural activities. This includes residues from crops such as maize, cassava and potatoes, and animal wastes.
2. Kenya has enough land to balance the future food demands of the growing population. However, a planting strategy needs to be in place to use 5% of available lands to produce sufficient lands to bal-

ance the future food demands of the growing population. There are large opportunities for greater utilisation of marginal and grasslands to produce >36 Mt of biomass material for briquette fuels, providing >8.01 TWh energy for Kenyan households and industry.
3. The utilisation of agricultural wastes and residues and dedicated production of biomass and energy can enable the country to achieve the climate and sustainability targets.
4. Policies required to promote sustainable production of biomass and agricultural waste within modern bioenergy systems will provide the country with low-carbon fuel options.

7.4 Challenges of Implementing the Paris Agreement and Key Lessons Learnt from the Kenyan Context

Kenya has shown leadership when it comes to climate issues, thanks to the support it has received from the Kenyan legislature coupled with the robust conduct and procedures considered to be one of the most progressive approaches around the continent. This is shown by how the Ministry has regular meetings on climate policy, budgetary allocations, and implementation (Westminster Foundation for Democracy (WFD) 2021). In as much as this is plausible, Kenya is still facing adaptation challenges evidenced by perception and deep uncertainty as to whether the climate change projections will materialise, which has led to the citizenry applying the "*wait and see approach*" as the rational decision, particularly at the grassroots level. In addition, the country also faces competing interests and other obstacles that make the operationalisation of the Paris Agreement challenging. These include:

A. **Competing investment interests**—Kenya, like many developing countries, has relatively insignificant carbon emissions and has hardly contributed to the climate change being experienced in the world. However, like most countries within the continent, Kenya currently stands at the crossroads of defining its energy future. The upscaling of renewable energy aligned with demand growth requires continuous additions in clean power capacity. Similarly, the nascent extractives sector is also a key factor in the investment horizon, given that the extractives sector also falls under one of the priority sectors towards the big four agenda, which is the development footprint under the incumbent regime. Kenya now needs to capitalise on its fossil fuel reserves while achieving a path consistent with the Paris Targets.

B. **Inadequate transmission lines capacity**—This includes the vulnerability of the energy system and increased generation from expensive fossil fuels in the dry months that occur due to the slight drop of renewable energy potential at this time.

7.5 Conclusion and Recommendations

Kenya has diversified its energy portfolio and boasts a robust renewable energy sector. In fact, Kenya is one of the top 5 Countries globally in the clean energy rankings, propelled mainly by policy incentives (donor-backed auction schemes and elaborate de-risking mechanisms. In addition, Kenya is the first sub-Saharan country to invest in the green bonds exchange to propel sustainable domestic investments. Although it has been able to make significant steps, there is still more to be done. These include:

I. Engaging in dialogues on local government and private sector collaboration to enhance sensitisation on climate action. There is a robust policy and framework towards climate action; however, it has not been fully implemented.

II. Promoting transparency and accountability in the Energy sector. Lack of openness hinders public trust over least-cost options and the future liabilities implied by take-or-pay obligations, particularly given sluggish demand growth. Currently, the country is undertaking a review of the current PPA's with a view to streamlining the electricity sector.

III. Kenya is already producing nearly 1 GW of power from geothermal sources thanks to the investment at the facilities in Olkaria near Lake Naivasha and the Menengai Crater near Nakuru. A relatively healthy system has evolved with state-owned exploration and private sector operators. Geothermal expansion holds tremendous potential in Kenya and its neighbours for this reliable zero-carbon power and industrial heat source (Kabeyi and Olarenwaju 2020).

IV. Encouraging and accelerating investments in Kenya's Green Bonds Programmes. So far, the uptake of the green bonds since its launch has been positive but relatively low due to the COVID-19 pandemic. There is thus a need to encourage and accelerate investments in Kenya's green bonds to facilitate low-carbon, climate-resilient growth and development, which could see Kenya achieve its NDC aspirations by 2030. In addition, this will ensure Kenya establishes itself as a cost-effective host country to GHG reduction projects.

V. Support local innovations and applications through institutional capacity-building programmes to propel local solutions for local problems. The country is home to one of the most established innovative ecosystems in Africa, with the majority of the innovations being developed by Kenya's youth, who are known to be vibrant and full of creativity. There is thus a great need to support youth-led innovations to support local initiatives such as the Bus Rapid Transit System in Nairobi, the non-motorised transport facilities, and the development of locally assembled electric vehicles.

VI. Develop industrial-level clean energy systems. There has been growing momentum on the potential of mini-grids, such as in the G20 Energy Access Action Plan for Africa. Locally initiatives such as the Kenya Off-Grid Solar Access Project (KOSAP) have also assisted in realising this potential. However, this should not just be limited to connecting a few households. The need for an industrial-level scale of this transformation should be given much more prominence, especially as the Big 4 Agenda has prioritised the manufacturing sector as the primary contributor to the growth of Kenya's GDP. Most of Kenya's private sector industry players including manufacturers still lobby for the need to have competitive tariff rates, which could definitely be derived from the cleaner energy sources being invested in at the moment. Some of the current large-scale manufacturers are mulling over installing their own clean energy power systems and altogether abandon the high tariffs they face from the main power supplier (Raballa 2021).

VII. Pioneer development of an energy transition strategy. Already Kenya is acting as the yardstick as well as a leader of excellence in renewable energy development (particularly in the wind, solar, and geothermal energy). It is thus poised to provide excellent thought leadership in pioneering efforts towards curating an energy transition strategy or roadmap that can be useful for the East African landscape and Africa as a whole.

VIII. Accelerate and capitalise on the cross-border electricity trade. This includes the flow of electricity occurring along the Kenya–Uganda interconnector. Kenya also has transmission interconnections with Ethiopia, thanks to the Lamu Port-South Sudan-Ethiopia-Transport (LAPSSET) Corridor project. Regional power systems can benefit from increased cross-border electricity trade, partly as a balancing mechanism against solar PV and wind power supply fluctuations. Although there has been a proposed framework in the form of the East Africa Power Pool, there is the need to accelerate this by ridding of the biggest obstacles to complete integration and functioning. Some of the obstacles that ought to be dealt with are the installation and completion of transmission facilities as well as enhancing the power generation capacity.

There is a need for significant investment in new generation and transmission grid infrastructure in Kenya and within the region. The huge investments that are needed must come from national and donor resources and private sector financing. Already Kenya tops in this area of private sector financing with swarms of power developers edging for a slot in the Kenya power sector. In addition, there is also a need to forge a real political commitment to regional solutions to energy supply (natural gas, oil, electricity) and to promote healthy competition in energy markets in the long term.

IX. Develop a strategy for recycling waste from clean energy sources (e.g. waste solar panels). Narratives of energy transitions have typically failed to address the environmental consequences of mass consumption. Locally and regionally, there is a heightened concern about managing E-waste. This growing concern needs to be attended to before it becomes a catastrophe. Currently, the pilot phase of tracking systems to ease the collection of products for recycling has been successful thanks to technology. However, there is a need to roll this out to a larger scale. In addition, significant investments need to be made into recycling e-waste as investments increase in solar and wind energy. Additionally, signing and operationalising the legislation around e-waste management can also go a long way.

References

IPCC. (2014). Working Group III Fifth Assessment Report of the Intergovernmental Panel on Climate Change Contribution to the Climate Change 2014. New York: Cambridge University Press. https://www.ipcc.ch/site/assets/uploads/2018/02/ipcc_wg3_ar5_full.pdf

KNBS, K. N. (2016). Economic Survey. Nairobi, Kenya. Retrieved from Kenya National Bureau of Statistics (KNBS): https://www.knbs.or.ke/?wpdmpro=economic-survey-2016

KNBS, K. N. (2019). Kenya Population and Housing Census Volume II. http://housingfinanceafrica.org/app/uploads/VOLUME-II-KPHC-2019.pdf.

PRB, Population Research Bureau. (2021). 2021 World Population Data Sheet. https://www.prb.org/news/2021-world-population-data-sheet-released/

GET.Invest. (2018). Renewable Energy Potential. https://www.get-invest.eu/marketinformation/kenya/renewable-energy-potential/

Gicheru, L. (2019). Policy Reform for Energy Transition in Kenya. https://www.worldenergy.org/news-views/entry/member-views-policyreform-for-energy-transition-in-kenya

Kenya's Updated Ntaionally Determined Contribution 2021https://www.iges.or.jp/sites/default/files/inlinefiles/8_Ressa_Kombi_Kenya%27s_updatd_nationally.pdf

Kahlen, L., Vivero, G. D., & Hecke., J. (2021). Kenya Power's Decarbonising the Energy Mix Initiative. New Climate Institute.

Kurdziel, M.-J., Day, T., Kahlen, L., & Schiefer., T. (2019). Retrieved from Climate change and sustainable development in the Kenyan electricity sector: Impacts of electricity sector development on Kenya's NDC. https://ambition-toaction.net/country-analysis-kenya/

Population Research Bureau. (2021). Population Indicators. https://www.prb.org/international/indicator/population-2050/snapshot/

Raballa, V. (2021, September 22nd). EABL to stop use of Kenya Power in Sh22bn plan. https://www.businessdailyafrica.com/bd/corporate/companies/eabl-to-stop-use-kenya-power-in-sh22bn-plan-3558336

Ritchie, H., & Roser, M. (2020). Energy Mix. Retrieved from Our World in Data: https://ourworldindata.org/energy-mix

Teye, J. (2018). Urbanisation and Migration in Africa. Centre for Migration Studies University of Ghana. Expert Group Meeting, United Nations Headquarters in New York. 1-2 November, 2018. New York: United Nations.

UN. (2014). World Urbanization Prospects The 2014 Revision. New York: United Nations.

UNFCCC. (2020, December 28th). Kenya's Updated NDC. https://www4.unfccc.int/sites/ndcstaging/PublishedDocuments/Kenya%20First/Kenya%27s%20First%20%20NDC%20(updated%20version).pdf

UNFCCC. (2021a). Interim NDC Registry. https://www4.unfccc.int/sites/ndcstaging/Pages/Home.aspx

UNFCCC. (2021b). The Paris Agreement. Retrieved from UNFCCC: https://unfccc.int/process-and-meetings/the-paris-agreement/the-paris-agreement

Kabeyi, M. J. B., & Olarenwaju, O. A. (2020). Viability of Wellhead Power Plants as Substitutes of Permanent Power Plants. Proceedings of the 2nd African International Conference, Harare, Zimbabwe, December 7–10, 2020. http://www.ieomsociety.org/harare2020/papers/77.pdf

Welfle, A., Chingaira, S., & Kassenov, A. (2020). Decarbonising Kenya's domestic & industry Sectors through bioenergy: An assessment of biomass resource potential & GHG performances. *Biomass and Bioenergy*, *142*, 105757. https://doi.org/10.1016/j.biombioe.2020.105757

Westminster Foundation for Democracy (WFD). (2021, May 5th). Stronger democratic process in Kenya to tackle climate change. https://www.wfd.org/2021/05/05/stronger-democratic-process-in-kenya-to-tackle-climate-change/

CHAPTER 8

The Political Economy of Decarbonising African Petro-cities: Governance Reconfigurations for the Future

Magnus C. Abraham-Dukuma, Okechukwu C. Aholu, Jesse Nyokabi, and Michael O. Dioha

Abstract The global decarbonisation agenda necessitates socio-technical and political economy recalibrations across the cities of the world. This chapter focuses on the political economy of decarbonising African

M. C. Abraham-Dukuma (✉)
Science and Strategy Hub, Gisborne District Council, Gisborne, New Zealand

O. C. Aholu
Nottingham Law School, School of Social Sciences,
Nottingham Trent University, Nottingham, UK

J. Nyokabi
Green Energy Pacesetter, Nairobi, Kenya

M. O. Dioha
Department of Global Ecology, Carnegie Institution for Science,
Stanford, CA, USA

S. I. Azubuike et al. (eds.), *Decarbonisation Pathways for African Cities*, Palgrave Studies in Climate Resilient Societies,
https://doi.org/10.1007/978-3-031-14006-8_8

petro-cities. Adopting a hybrid qualitative approach consisting of theoretical exploration and content analysis, we deconstruct the political economy of decarbonisation in African petro-cities and explore possible governance measures that promote decarbonisation while supporting economic prosperity. Our policy recommendations include recalibrating both supply and demand sides of energy culture, a shift in the economic dynamics of petro-dependency and rent-seeking, governance reforms to promote clean energy democracy and energy decentralisation, incentivising active private sector engagements, and adopting an expert approach to decarbonisation within specific geopolitical contexts.

Keywords Decarbonisation • African petro-cities • Oil and gas • Energy transition • Political economy

8.1 Introduction

Rapid transitions in virtually all facets of the global system are necessary to save planetary life forms from the impending consequences of a changing climate. Consequently, there are numerous governance, economic/financial, institutional, behavioural, and systematic implications for achieving the climate targets set under the Paris Agreement. These also have ramifications for the world at a macro-level and states and cities at a micro-level. In this chapter, our focus is on the micro-level in relation to cities, due to the prominent position of cities in population growth, possible increase in both urban industrial activities, and the associated greenhouse gas (GHG) emissions. Evidence from the urban growth literature shows that cities will continue to grow in population, even though some may grow rapidly while others grow slowly (Duranton and Puga 2013). Research shows that the global population would reach 9.7 billion in 2050, with 68 per cent of this number expected to live in urban areas (essentially cities), with the resultant effect of increasing emissions (UNDESA 2019).

Relatedly, a significant implication that has become the central subject of disciplinary and multidisciplinary scholarship is the need to decarbonise industrial processes and economic activities globally (Rosenbloom and Rinscheid 2020; Wimbadi 2020). However, there are numerous parallels and political economy trade-offs regarding decarbonisation within global, national, and sub-national spheres. It is also noteworthy that about 137 countries have set a carbon neutrality target for various years—for example, 2030 for Uruguay, 2035 for Finland, 2040 for Austria and Iceland, 2045 for Germany and Sweden, and 2050 for Canada, Denmark, France, and many others (Wallach 2021).

As brilliant as the idea of decarbonisation sounds, it is also susceptible to the intricacies of national politics and the prevalent political economy matrix of different countries and cities (Alexander 2020; Bernstein 2018). This explains the intractable nexus between economic development and carbon-intensive industries in different countries, thereby stalling decarbonisation efforts globally (Wesseling et al. 2017). This fact underpins the emerging and growing concept of socio-technical transitions, requiring a recalibration of socio-technical systems along the path of sustainable development (Hess and Sovacool 2020; Sovacool et al. 2020). Consequently, there is a need to seek a logical path towards understanding how to decouple economic development from carbon-intensive industries. While most studies have focussed on political economy at the national scale (Barker and Crawford-Brown 2014; Hildingsson 2019; Lachapelle et al. 2017), here we intend to follow a bottom-up approach to study the problem by focussing on cities, as local actions are necessary to address the current climate challenge (Salon et al. 2014).

Against this background, this chapter seeks to explain the political economy of decarbonisation in African petro-cities. It attempts to provide logical governance reconfigurations that foster decarbonisation and support future economic prosperity. The chapter adopts a hybrid of theoretical exploration and content analysis. The political economy of decarbonisation and the emerging concept of socio-technical transition underpin the theoretical framework of this chapter. We then apply the framework to the African context, drawing from the relevant literature, to understand the present state of knowledge regarding the political economy of decarbonisation in Africa. Then, we adopt content analysis to examine current governance trajectories in African petro-cities as a premise for exploring optimal governance measures that support decarbonisation. We have focused predominantly on four petro-cities in Africa's prominent petroleum-producing jurisdictions—Hassi Messaoud (Algeria), Luanda (Angola), Tripoli (Libya), and Port Harcourt (Nigeria). We now proceed to set a theoretical context in the next section.

8.2 Theoretical Context

We explore 'socio-technical transition' and 'political economy of decarbonisation' as the theoretical foundations of our analysis. While they do not necessarily represent an exclusive framework for the themes discussed in the chapter, we believe that they significantly capture the idea of decarbonisation at both general and macro levels; and the decarbonisation of African petro-cities at a micro level.

The growing concept of socio-technical transition broadly encapsulates systemic and socio-technical shifts consistent with sustainable development and decarbonisation (Hess and Sovacool 2020; Köhler et al. 2019; Markard et al. 2012). This implies transformations in virtually all facets of human endeavours—for example, energy use (for transportation, heating, and electricity), agriculture, and various social systems and modes of business (Geels et al. 2017). For energy use, it is imperative to transition to renewables. For agriculture and business systems, the world needs to transition to regenerative modes of agriculture and sustainable business models that support the global climate agenda. There are certainly numerous spheres of human endeavour and social systems where transition can occur, and we cannot exhaustively cover them in this book chapter.

However, we construe petro-cities as part of socio-technical systems that need rapid multifaceted shifts that drive decarbonisation objectives. Nonetheless, it is evident that socio-technical transitions (including decarbonisation objectives and the general theme of energy transition) occur amidst multi-dimensional trade-offs, which countries and cities should consider in planning and policy processes (Nasirov et al. 2020). For example, it is important to understand and plan strategies for decoupling economic development from carbon-intensive activities and industries (Deutch 2017). Within the context of African petro-cities, political economy trade-offs correlate to decarbonisation and energy transition trajectories. This consequently justifies the logic to also explore 'the political economy of decarbonisation' as a theoretical foundation to underpin the discussion in the chapter.

Drawing from a rich repository of scholarship, we define the political economy of decarbonisation as the multifaceted dynamic relationships and trade-offs that surface in the clean transition journey of geographical and geopolitical entities at international, national, municipal, and city levels (Alexander 2020; Biber 2017; Lachapelle et al. 2017; Pearse 2020). For example, huge economic dependencies and governmental rents from

Table 8.1 Economic impact of oil on African OPEC member countries

Country	*Oil contribution to GDP in 2019*	*Oil contribution to export revenue in 2019*
Algeria	20 per cent	85 per cent
Angola	50 per cent	89 per cent
Libya	60 per cent	69 per cent
Nigeria	10 per cent	85 per cent

Source: Adapted by Authors from the OPEC Annual Statistical Bulletin 2020 (OPEC 2020)

carbon-intensive industries (with the lack of viable alternatives) represent striking political economy factors that hinder strong climate action (Gaulin and Le Billon 2020; Lamb and Minx 2020; Röttgers and Anderson 2018). We apply these theoretical frames to African petro-cities in the subsequent sections of the chapter.

8.3 The Political Economy of African Petro-cities

The petroleum industry, in a stable fiscal climate, is one of enormous complexity and reward. As an industry central to security, prosperity, and the very nature of civilisation, it is usual for petroleum extracting countries to accelerate the process of profiting from the resource. As Table 8.1 shows, oil contributes significantly to the gross domestic product (GDP) and export revenue of Algeria, Angola, Libya, and Nigeria—all member countries of the Organization of the Petroleum Exporting Countries (OPEC). For this assertion, we note that we have used data from 2019, being the most available recent data for our chosen parameters—oil contribution to gross domestic product (GDP) and oil contribution to export revenue—for the four countries at the time of writing.

As it becomes more necessary to constrain the production of petroleum due to climate change concerns, one fundamental question that has been asked is whether OPEC is on the path of demise due to the inextricable dependence of its member countries on the oil economy (van de Graaf 2017). This question may also be asked about petro-cities within the selected African OPEC member countries. Within the micro-level dimension of our study, we interrogate the prevalent political economy of the selected African petro-cities by identifying their dominant revenue streams (present and future) and inquiring into the existence or not of their climate action policy/plan and energy transition policy/plan. We summarise

Table 8.2 Dynamics of African petro-cities

Country	*Cities*	*Major revenue stream*	*Future revenue stream plan*	*Climate action policy/plan*	*Energy transition policy/plan*
Algeria	Hassi Messaoud	Almost total economic reliance on the continued existence of the oil and gas industry	No known policy/plan	No known policy/plan	No known policy/plan
Angola	Luanda	Huge dependence on oil export revenues	No known policy/plan	No known policy/plan	No known policy/plan
Libya	Tripoli	Hugely reliant on oil revenue	No known policy/plan	No known policy/plan	No known policy/plan
Nigeria	Port Harcourt	Allocation from national oil revenue and internally generated revenues through taxes and levies	No known policy/plan	No known policy/plan	No known policy/plan

these as the dynamics of the case study of African petro-cities in Table 8.2. We quickly observe that it was difficult to get primary data on the selected cities and had to rely heavily on secondary sources.

In Algeria, the oil city of Hassi Messaoud—located 85 kilometres south-east of the Ouargla Province—is essentially a service town for surrounding oil and gas operations and workers, with subservient businesses (Ham et al. 2007). Thus, economic activities in the city hugely rely on the continued existence of the oil and gas industry to thrive. There is no known future revenue stream plan, climate action policy/plan, or energy transition policy/plan for the city. In Angola, there is also a huge dependence on oil revenue from the national level to city level. As Table 8.1 shows, approximately 90 per cent of the national export earnings come from the oil and gas industry. This also has ripple economic effects on Luanda, being the country's capital and major city (Tvedten et al. 2018). Yet urban poverty, corruption, and inflation are mainstream indices of the resource curse paradox in the city (and by extension, the nation) (Tvedten et al. 2018). Our search did not reveal any known city-centric future revenue stream plan, climate action policy/plan, or energy transition policy/plan.

Amongst the four African OPEC member countries, Libya has the highest oil contribution to GDP (60 per cent) (Table 8.1). This tells much about the sustenance of its capital and largest city—Tripoli. There is no known future revenue stream plan, climate action policy/plan, or energy transition policy/plan. Port Harcourt is fondly known as the oil capital of Nigeria. Its government fiscal records are not within the public domain. However, it is common knowledge that a large chunk of the city's revenue comes from monthly federal fiscal allocation derived from oil revenue as well as internally generated revenues (principally city/local government levies and taxes). Within the prevailing revenue allocation formula in Nigeria, over 50 per cent of oil revenues go to the federal government, 26 per cent to states and 20 per cent to local governments, as well as 13 per cent derivation to oil-producing states (Onuigbo and Innocent 2015). Small and medium-sized businesses also hugely rely on the patronage of the workforce of international oil (and oil servicing) companies within the city and around its outskirts. To our knowledge, there is no known future revenue stream plan for the city independent of petroleum revenues, climate action policy/plan, or energy transition policy/plan.

The oil and gas sector occupies a dominant position in the economy of the focus cities. In addition, the absence of alternative future sources of revenue creates a carbon-constrained political economy at both national and city levels. More significantly, pursuing a decarbonisation agenda without a future sustainable revenue source will facilitate the loss of income from the incumbent oil sector in these cities, thereby exposing them to economic distress and security challenges. These can be summarised as three general barriers to decarbonisation in these petro-cities—non-diversification of present revenue streams, reliance on traditional carbon-intensive industries, and lack of ambitious governance regimes.

Furthermore, there are several reasons for the want of energy transition and climate policies/plans in the focus petro-cities. The first is the lack of data on the sources of emissions in African cities generally. Cities need data to plan emissions reduction policies and deliver development priorities and climate goals (World Bank 2016). The second reason is a general lack of institutional awareness of what needs to be done in the context of city planning. This problem is compounded by the over-politicisation of governance in cities, a factor that makes it difficult to sustain public participation that helps access local experts and relevant stakeholders for policy planning. The third is the centralisation of energy governance. The energy topography and the provision of electricity in the continent are more characterised by a centralised, large scale and one-directional model. This concentrated model of energy governance suggests that decarbonisation is mainly a national obligation with no city inputs. Contrariwise, local authorities and cities will need to play fundamental roles in advancing the global climate change mitigation agenda (OECD 2014).

8.4 Future Scenarios

Here, we identify five projections as probable future scenarios for African petro-cities, drawing from the foregoing analysis, the prevailing logic of energy transition, and the global decarbonisation agenda. The striking projections that easily lend themselves to our qualitative reasoning encompass energy and political economy themes, which we set forth in the following paragraphs and conceptualise through a hierarchical pyramid prism (Fig. 8.1).

8.4.1 Market-Deficit Petro-cities

The international political economy of oil and gas is constantly driven by competition among petroleum-producing jurisdictions for market shares

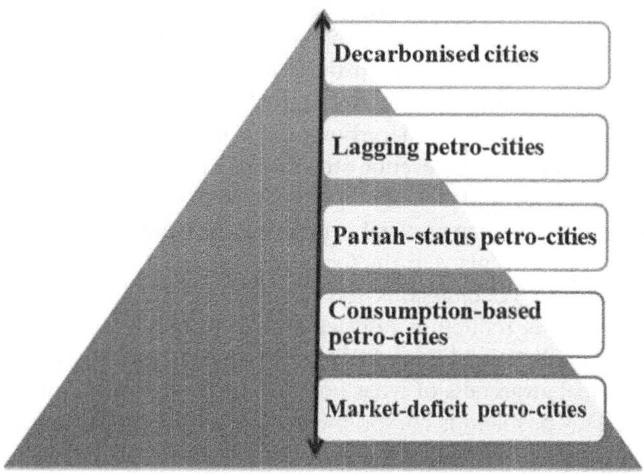

Fig. 8.1 Future scenarios hierarchical pyramid for African petro-cities. Source: Authors

and cooperation to assure energy security (Bolanos 2018). These twin factors validly constitute some major rationales for OPEC's establishment and continued existence. Relatedly, the global decarbonisation agenda will further exacerbate the volatility of already shrinking global oil markets. The implication for petro-cities is a deficit in markets for trading petroleum products, which consequently leads to over-supply of products, further fall in oil prices, and a consumption-based economy for petro-cities that fail to diversify both economic and energy sources. We place petro-cities in these circumstances at the bottom position of Fig. 8.1.

8.4.2 *Consumption-based Petro-cities*

A potential consequence of the changing petroleum market dynamics is that oil-producing countries and petro-cities will be consumption-based instead of deriving revenues from international petroleum trade. The logic is simple. Amidst global shrinking oil markets and likely continuity of petroleum production in African petro-cities, there will be less international supply (product sale). This can be worse if these countries and cities lack ambitious and feasible economic diversification plans. However, as will be seen subsequently in this chapter, there are good examples of petro-cities with good plans to transition from a petroleum-dependent economy to sustainable business and economic models that align with a low-carbon future.

8.4.3 *Pariah-status Petro-cities*

In the middle of the hierarchical pyramid for African petro-cities is 'pariah-status petro-cities'. Given the characteristics of positions 1 and 2 of the pyramid, African petro-cities will need to reconfigure their socio-technical systems, especially the economic base, and energy sources to remain relevant in relation to the wider global village. In the absence of such reconfigurations, African petro-cities—and most likely other petro-cities under the OPEC umbrella and around the world—face the risk of being left behind in the scheme of things with respect to the ongoing global transitions. The world could progress towards economic models and smart energy systems for the future, while incumbent petro-cities that refuse to transition could suffer detrimental disconnections from the global system. This bleak possibility necessitates serious and urgent governance interventions to drive socio-technical transitions and economic diversification.

8.4.4 *Lagging Petro-cities*

As the energy transition discussion continues, an emerging and evolving logic is for developing states in Africa with huge dependencies on fossil fuels to progressively develop clean energy alternatives at a slower pace in comparison with their developed country counterparts. This is what has emerged as the 'energy progression dialogue' (Nalule 2021). This can have implications at the micro-level for African petro-cities, where they could pursue decarbonisation and economic diversification objectives at a slower pace. This pursuit moves them from the positions of market-deficit, consumerist- and pariah-status petro-cities to a position where they slowly move towards energy transition and decarbonisation. Since they will not move at the same pace as their developed counterparts, we categorise them in the hierarchical pyramid for African petro-cities (Fig. 8.1) as 'lagging petro-cities'.

8.4.5 *Decarbonised Cities*

A desirable possibility for African petro-cities will be to attain complete or near-complete decarbonisation. Generally, there are uncertainties surrounding the temporal dynamics of decarbonisation and energy transition goals (Sovacool 2016). Without prejudice to prognoses and conjectures by scholars and relevant organisations, it is difficult to predict when countries and cities will achieve comprehensive decarbonisation. However, as the global decarbonisation agenda continues, and as African petro-cities undertake energy progression, it is safe to express optimism about future possibilities for these cities to attain the status of 'decarbonised cities', being the fifth position of the hierarchical pyramid for African petro-cities (Fig. 8.1). Whichever the case, a lot depends on how these cities initiate governance and socio-technical transition measures that support decarbonisation. We now turn to this point in the next section.

8.5 Governance and Socio-technical Reconfigurations

Based on the theoretical context set out in Sect. 8.2 above, we argue that it is possible for African petro-cities to drive governance and socio-technical reconfigurations that support decarbonisation objectives. From global conceptual and empirical dimensions, research provides a solid

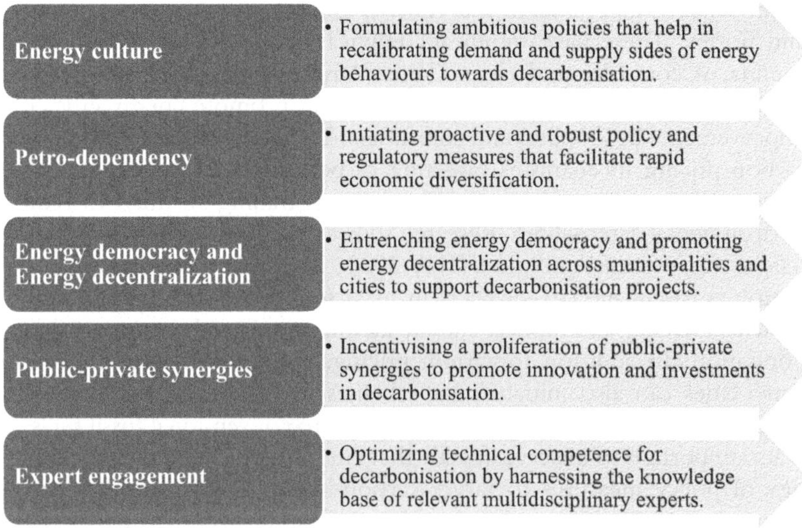

Fig. 8.2 Governance reconfigurations for decarbonising African petro-cities. Source: Authors

evidence base for the feasibility of policy measures that incentivise a reduction of GHG emissions (Barker and Crawford-Brown 2014; Dioha et al. 2020; Dioha and Kumar 2020). We draw from this established foundation to explore governance measures within an African context, encompassing suitable future directions that promote decarbonisation in petro-cities, including but not limited to the following: optimising energy culture; addressing petro-dependency; promoting municipal energy democracy and energy decentralisation; intensifying public-private synergies and expert engagement. We provide a graphic summary (Fig. 8.2) and a succinct discussion of these themes in the next paragraphs.

8.5.1 *Fixing Demand-Supply Sides of Energy Culture*

Energy culture presupposes the inter-related interactions that exist between cognitive norms (e.g., beliefs, understandings), material culture (e.g., technologies, building form), and energy practices (e.g., activities, processes) (Bleischwitz et al. 2010; Stephenson et al. 2010, 2015). This can be understood from two broad dimensions—supply-side (energy supply chains) and demand-side (energy consumption patterns) practices and norms. These have significant implications for the decarbonisation agenda of countries and cities. Thus, African petro-cities will need to entrench regulatory and policy measures that promote energy efficiency improvements by suppliers and consumers. These can include well-tailored carbon pricing mechanisms targeting carbon-intensive industries within cities (carbon taxes and functional emissions trading systems) to reduce environmental externalities. Research shows that these have proven potentials to drive decarbonisation (IEA 2021; Metcalf 2019). Another vital option is to engage city dwellers in mass sensitisation on energy home-centric efficiency and modest climate action measures that support decarbonisation. In addition to carbon pricing, the governments of African petro-cities can also initiate pilot schemes to incentivise the uptake of renewables for energy services in preference to conventional fossil fuels. It is also important to observe quickly that cities may be limited to the typology of policy measures to target carbon-intensive industries, especially where the jurisdiction to govern such industries (especially the petroleum industry) lies within the purview of the federal government in a federal system of government. In such circumstances, petro-cities can pursue collaboration with the central government to roll out suitable policy measures.

8.5.2 Recalibrating Petro-dependency and Rent-Seeking

A known argument in literature for addressing petro-dependency is economic diversification (Alsharif et al. 2017; Olander 2019). As a fact, incidences of resource curse and the prevalent weak regulatory regimes targeting carbon-intensive industries across Africa at a macro-level are all connected to intractable rent-seeking and economic dependency on the petroleum industry (Lamb and Minx 2020; Okpaleke and Abraham-Dukuma 2020). Since decarbonisation and transition to low-carbon energy sources demand that majority of the world's fossil fuel supplies cannot be burnt (Mcglade and Ekins 2014), it is imperative that fossil fuel-reliant cities in Africa diversify their economies to other sustainable revenue sources. In complementing national economic transformation measures, these cities need to initiate sustainable city-centric business models that are independent of the oil and gas industry and other carbon-intensive industries. For example, it may be possible to concentrate on developing non-oil revenue sources such as the agricultural and hospitality industries. This is the logic underpinning the emergence of innovative economic/business models for a sustainable low-carbon future (Markard 2018; Newell 2020).

8.5.3 Promoting Energy Democracy and Energy Decentralisation

Energy democracy is one of the approaches for achieving energy transition and decarbonisation objectives. A uniform definition of the concept is yet to crystallise, but from a review of the relevant literature, it can be described as a collective term capturing the merger of technological energy transition and the strengthening of democracy as a practice and public participation in energy decision-making (Szulecki 2018; Szulecki and Overland 2020; van Veelen and van der Horst 2018). Equally relevant is the related concept of energy decentralisation, which simply means to generate energy at different locations where it is needed rather than generating it through a central main grid and distributing across the state or nation. Energy sources generated through a decentralised energy system can range from micro-renewables to clean cooling and heating systems. Energy democracy and a decentralised energy regime can afford communities and cities an opportunity to invest in city- and community-driven decentralised renewable energy technologies to create 'renewable energy communities' (Heldeweg and Séverine Saintier 2020).

At a general level, energy decentralisation is becoming more popular as a fundamental requirement for addressing energy access issues across sub-Saharan Africa (Zalengera et al. 2020). Research shows that states can formulate policy instruments that promote energy democracy and energy decentralisation (Burke and Stephens 2017). However, this depends on the regulatory framework for energy generation and distribution at national levels and on a case-by-case basis. Germany, the United States of America, and Denmark represent good examples where energy democracy and energy decentralisation have thrived, thereby allowing cities and municipalities to invest in decentralised renewable energy projects to drive decarbonisation goals (Mendonça et al. 2009; Paul 2018). Across Africa, the regulation of energy services (generation and distribution) is predominantly centralised (Zalengera et al. 2020). In some cases, there is decentralisation of energy generation and centralisation of energy transmission through a single national grid. Nigeria exemplifies this fact. This needs to change so that cities will be able to facilitate off-grid energy generation and distribution by harnessing decentralised energy technologies.

However, energy democracy and energy decentralisation may not thrive in non-democratic countries and cities (Delina 2018). It is a recognised fact that the governance culture of a country influences the socio-political narrative or representation of that country (Mehmetaj 2014). It would be an error to all the features of a democratic system in a nation under a military rule. So, if a draconian system of governance operates in a country, it is difficult to canvass democracy in any form. This is applicable to 'energy democracy'. Within the context of African petro-cities, the successful implementation of energy democracy and energy decentralisation will largely depend on the quality and legitimacy of political systems from the central to decentralised spheres of governance. Therefore, it is more probable to have functional energy democracy leading to decentralised energy systems that support decarbonisation in countries with strong democratic norms and institutions; and less likely to have such experience in draconian and despotic governance regimes or regimes with poor democratic culture.

8.5.4 Active Private Sector Engagements (Investments and Innovation)

As the governance of decarbonisation evolves, it is becoming more pertinent to acknowledge the crucial role that non-state actors play in terms of technical competence, technological innovation, and investments (Alkhani 2020). The local governments of African petro-cities and municipalities can initiate suitable incentives to engage the private sector (essentially carbon-intensive industries and other corporate entities) to engage in suitable decarbonisation projects through public-private partnerships. This can be an essential component of their long-term roadmaps, covering economy-wide and energy-centric considerations that support a low-carbon future.

8.5.5 Adopting an Expert Approach to Decarbonisation

Research reveals the necessity of underpinning state climate policy and decarbonisation measures on a robust scientific foundation and the usefulness of harnessing and developing relevant competencies for such evidence base (Abraham-Dukuma et al. 2020a; Dioha et al. 2020). Drawing on this premise, African petro-cities may profit from adopting an expert approach to decarbonisation, requiring harnessing the expertise of multiple experts to provide relevant advisory services on an ongoing basis. A good example is to set up a city-centric task force or commission on decarbonisation. Such an entity could provide expert advisory services on the different dimensions and numerous implications of decarbonisation and economic diversification. It is possible that these processes could inform the formulation of suitable policies and governance measures to achieve decarbonisation objectives across African petro-cities.

8.6 Learning from Three Good Examples

Beyond the frontiers of Africa, there are good examples of petro-cities with ambitious plans for decarbonisation. We present three such examples in this section of the chapter—the Abu Dhabi Economic Vision 2030, the Riyadh Vision 2030, and the Taranaki 2050 Roadmap.

8.6.1 The Abu Dhabi and Riyadh 2030 Visions

Riyadh and Abu Dhabi are two prominent petro-cities in two Middle Eastern countries—Saudi Arabia and the United Arab Emirates (U.A.E.)—, and highly reputable producer economies with huge economic dependency on the petroleum industry. Owing to concerns about climate change and the numerous uncertainties surrounding the future of oil and gas resources, these two cities have initiated ambitious energy transition and economic diversification plans by 2030. For Riyadh (Saudi Arabia), the vision is to ensure economic diversification and reduce the dependence on petroleum revenues by 2030 (The Kingdom of Saudi Arabia 2020). It proposes to increase the quantum of revenue from non-oil exports from 16 per cent to 50 per cent and increase non-petroleum industry revenue from 163 billion to 1 trillion Saudi Riyal by the year 2030. There are also renewables investment strategies in the plan, essentially to recalibrate the energy supply dynamics in favour of clean energy sources.

The Government of Abu Dhabi (in the UAE) has a similar diversification plan with two dominant policy priorities. One is to build a sustainable economy and the other is to ensure a balanced social and regional economic development approach that benefits all citizens (The Abu Dhabi Government 2020). These policy priorities rest on the idea of decoupling economic development from the oil and gas industry in ways to assure future prosperity. Thus, the plan also underscores promoting a low-carbon energy economy and focuses on seven strategic areas. These include building an efficient and globally integrated business environment; adopting fiscal measures that both avert inflation and promote sustainability; establishing resilient monetary and financial markets; driving energy efficiency improvements; developing sufficient supportive infrastructure to support sustainable business models for economic growth; developing highly skilled productive workforce; and incentivising fiscal supports from key financial markets.

8.6.2 *The Taranaki 2050 Roadmap*

A turn to the Taranaki Region of New Zealand also reveals exemplary reconfigurations that support decarbonisation. New Zealand has 17 sedimentary basins, but petroleum exploration and production activities occur only in 1 basin in the Taranaki region (PEPANZ 2019). Therefore, the region is the hub of the New Zealand oil and gas industry and comprises eight urban areas/cities—Waitara, Inglewood, New Plymouth, Patea, Hāwera, Stratford, Opunake, and Eltham. The economies of these cities hugely rely on petroleum revenues in the forms of taxes, royalties, and employment generation, hence highly vulnerable to the economic shocks and effects of the new policy measure by the New Zealand government to restrict offshore petroleum activities in the region (Abraham-Dukuma et al. 2020b; Kubiak et al. 2019).

Cognizant of the region's susceptibility to the uncertainties of the petro-economy and the need to usher economic and energy transitions, the regional government of Taranaki introduced the Taranaki 2050 Roadmap, which sets out to achieve net-zero emissions and incentivise sustainable economic growth (Venture Taranaki 2019). To achieve these policy goals, it introduces transition pathways that require investments in low-carbon energy sources, economic diversification, and multi-sectoral growth across the length and breadth of the region's cities.

There can be further discussion regarding the Taranaki 2050 Roadmap, and the 2030 visions of Riyadh and Abu Dhabi, as well as numerous other petro-cities and regions of the world; but the central point is the insight to draw from these strategies.

8.7 Conclusion

This chapter set out to provide an appreciable understanding of the political economy of decarbonisation in African petro-cities to proffer potential governance measures that promote decarbonisation while supporting economic prosperity. This has been conducted within the theoretical frame of socio-technical transition and the political economy of decarbonisation. It is evident that petro-cities are part of the socio-technical systems in need of rapid transformation, including economic and energy models that support a low-carbon future. From the analysis, the economies of major African petro-cities are hugely reliant on the fossil fuel industry. This will need to change through serious economic diversification measures.

We have also discussed five possible future scenarios for African petro-cities. First, petro-cities that fail to diversify their economic base in line with the energy transition are likely to have shrinking markets in the future as the world shifts carbon-intensive fuels to renewables, thus creating what we have referred to as 'market-deficit petro-cities'. The logical consequence of this is the prevalence of consumption-based petro-cities that will earn little or no foreign revenues from international oil and gas sales. The corollary consequences of these scenarios are that petro-cities also risk lagging in the global energy transition and having a pariah status in the scheme of international energy politics. However, we have not painted an all-morbid picture. Africa's petro-cities can still be decarbonised in ways that economic development is decoupled from carbon-intensive activities. This is the fifth scenario we discussed in this chapter, but it will require serious planning and a strategic direction that looks to the future. We have illustrated this point using good examples of other petroleum-dependent economies that have defined a roadmap for both energy transition and economic diversification from a fossil-dominated economy to sustainable business models to support a low-carbon future. The Abu Dhabi and Riyadh Visions 2030, and the Taranaki 2050 Roadmap are specific examples discussed in this chapter.

Further, to facilitate serious local-level decarbonisation efforts, municipal, regional and city governments will need to adopt a futuristic economic lens to decarbonisation with well-considered plans that capture various aspects and trade-offs associated with decarbonisation. We have also suggested specific areas that petro-cities can focus strategic attention on in the process of reconfiguring their systems for the future. First, they will need to carefully plan how to fix the demand and supply sides of energy culture and how the two sides support a low-carbon future. Second, rent-seeking political systems will need to recalibrate their approach to governance and focus more on innovative revenue generation avenues other than hugely relying on finite oil revenues. Third, energy democracy and energy decentralisation will be pivotal in helping cities chart a sustainable future, but this would largely depend on the prevalent national political culture in various nations. Fourth, petro-cities can leverage private sector financing for sustainable investments using well-designed incentives. Lastly, the policy and strategic directions of petro-cities can profit immensely from an expert advisory team that has to be carefully sourced. In recent times, it is also becoming more reasonable for countries to drive their climate action using experts drawn from various spheres and disciplines pertaining to climate change.

We note our predominant reliance on secondary sources in extrapolating the political economy circumstances across the case study cities due to the glaring lack of primary data. We acknowledge this as one of the limitations of our study. The second limitation is the lack of in-depth analysis on some of the themes explored due to the limited scope of this book chapter. Based on the future availability of primary data, there could be further detailed and more insightful analysis on the topic. However, we believe that our modest effort in the present inquiry helps to inform governance and socio-technical reconfigurations on the subject matter.

References

Abraham-Dukuma, M. C., Dioha, M. O., Bogado, N., Butu, H. M., Okpaleke, F. N., Hasan, Q. M., Epe, S. B., et al. (2020a). Multidisciplinary Composition of Climate Change Commissions: Transnational Trends and Expert Perspectives. *Sustainability 2020, Vol. 12, Page 10280*, *12*(24), 10280. Multidisciplinary Digital Publishing Institute. Retrieved October 5, 2021, from https://www.mdpi.com/2071-1050/12/24/10280/htm

Abraham-Dukuma, M. C., Okpaleke, F. N., Hasan, Q. M., & Dioha, M. O. (2020b). The Limits of the Offshore Oil Exploration Ban and Agricultural Sector Deal to Reduce Emissions in New Zealand. *Carbon & Climate Law Review*, *14*(2), 107–117. Lexxion Publisher. Retrieved August 1, 2020, from https://cclr.lexxion.eu/article/CCLR/2020/2/6

Alexander, S. (2020). The Political Economy of Deep Decarbonization: Tradable Energy Quotas for Energy Descent Futures. *Energies (Basel)*, *13*(17), 4304-.

Alkhani, R. (2020). Understanding private-sector engagement in sustainable urban development and delivering the climate agenda in northwestern Europe—a case study of London and Copenhagen. *Sustainability (Switzerland)*, *12*(20).

Alsharif, N., Bhattacharyya, S., & Intartaglia, M. (2017). Economic diversification in resource rich countries: History, state of knowledge and research agenda. *Resources Policy*, *52*.

Barker, T. B., & Crawford-Brown, D. (2014). *Decarbonising The World's Economy: Assessing The Feasibility Of Policies To Reduce Greenhouse Gas Emissions.* (T. B. Barker & D. Crawford-Brown, Eds.) *Decarbonising The World's Economy: Assessing The Feasibility Of Policies To Reduce Greenhouse Gas Emissions.* World Scientific Publishing Company.

Bernstein, S. (2018). The politics of decarbonisation and the catalytic impact of subnational climate experiments. *Policy sciences, 51*(2), 189–211.

Biber, E. (2017). The political economy of decarbonisation: a research agenda. *Brooklyn law review, 82*(2), 605-.

Bleischwitz, R., Bader, N., Stephenson, J., Barton, B., Carrington, G., Gnoth, D., Lawson, R., et al. (2010). Energy cultures: A framework for understanding energy behaviours. *Energy Policy, 38*(10).

Bolanos, J. A. (2018). The Political Economy of Oil: The Limits of Our Ability to Prevent Failure. In S. Raszewski (Ed.), *The International Political Economy of Oil and Gas* (pp. 57–67). Palgrave Macmillan.

Burke, M. J., & Stephens, J. C. (2017). Energy democracy: Goals and policy instruments for sociotechnical transitions. *Energy Research & Social Science, 33*, 35–48.

Delina, L. L. (2018). Can energy democracy thrive in a non-democracy? *Frontiers in Environmental Science, 6*(JAN).

Deutch, J. (2017). Decoupling Economic Growth and Carbon Emissions. *Joule, 1*(1), 3–5. Cell Press.

Dioha, M. O., Abraham-Dukuma, M. C., Bogado, N., & Okpaleke, F. N. (2020). Supporting climate policy with effective energy modelling: A perspective on the varying technical capacity of South Africa, China, Germany and the United States. *Energy Research & Social Science, 69*, 101759. Retrieved from https://doi.org/10.1016/j.erss.2020.101759

Dioha, M. O., & Kumar, A. (2020). Exploring greenhouse gas mitigation strategies for agriculture in Africa: The case of Nigeria. *Ambio, 49*(9).

Duranton, G., & Puga, D. (2013). The growth of cities. In P. Aghion & D. Steven (Eds.), *Handbook of Economic Growth* (pp. 781–1106). Elsevier.

Gaulin, N., & Le Billon, P. (2020). Climate change and fossil fuel production cuts: assessing global supply-side constraints and policy implications. *Climate Policy*, 1–14. Taylor and Francis Ltd.

Geels, F. W., Sovacool, B. K., Schwanen, T., & Sorrell, S. (2017). The Socio-Technical Dynamics of Low-Carbon Transitions. *Joule, 1*(3), 463–479. Cell Press.

van de Graaf, T. (2017). Is OPEC dead? Oil exporters, the Paris agreement and the transition to a post-carbon world. *Energy Research and Social Science, 23*, 182–188.

Ham, A., Luckham, N., & Sattin, A. (2007). *Algeria*. London: Lonely Planet.

Heldeweg, M. A., & Séverine Saintier. (2020). Renewable energy communities as 'socio-legal institutions': A normative frame for energy decentralisation? *Renewable and Sustainable Energy Reviews, 119.*

Hess, D. J., & Sovacool, B. K. (2020). Sociotechnical matters: Reviewing and integrating science and technology studies with energy social science. *Energy Research & Social Science*, *65*, 101462.

Hildingsson, R. (2019). The green state and industrial decarbonisation. *Environmental politics*, *28*(5), 909–928.

IEA. (2021, August 4). Carbon pricing can help put clean energy at the heart of stimulus packages. *IEA*. Retrieved October 3, 2021, from https://www.iea.org/commentaries/carbon-pricing-can-help-put-clean-energy-at-the-heart-of-stimulus-packages

Köhler, J., Geels, F. W., Kern, F., Markard, J., Onsongo, E., Wieczorek, A., Alkemade, F., et al. (2019). An agenda for sustainability transitions research: State of the art and future directions. *Environmental Innovation and Societal Transitions*, *31*, 1–32.

Kubiak, L., Isack, E., & Welvaert, M. (2019). *Economic impact of ending new oil and gas exploration permits outside onshore Taranaki: A regional CGE analysis NZIER report to PEPANZ.* Retrieved January 4, 2021, from www.nzier.org.nz.

Lachapelle, E., MacNeil, R., & Paterson, M. (2017). The political economy of decarbonisation: from green energy 'race' to green 'division of labour.' *New Political Economy*, *22*(3).

Lamb, W. F., & Minx, J. C. (2020). The political economy of national climate policy: Architectures of constraint and a typology of countries. *Energy Research & Social Science*, *64*, 101429. Retrieved July 21, 2020, from https://doi.org/10.1016/j.erss.2020.101429

Markard, J. (2018). The next phase of the energy transition and its implications for research and policy. *Nature Energy*, *3*(8).

Markard, J., Raven, R., & Truffer, B. (2012). Sustainability transitions: An emerging field of research and its prospects. *Research Policy*, *41*(6), 955–967. North-Holland.

Mcglade, C., & Ekins, P. (2014). Un-burnable oil: An examination of oil resource utilisation in a decarbonised energy system. *Energy Policy*, *64*, 102–112. Retrieved July 19, 2020, from https://doi.org/10.1016/j.enpol.2013.09.042

Mehmetaj, J. (2014). The Impact of the Political Culture in Political System and Rule of Law: Albana Case. *European Scientific Journal*, *1*, 483–490.

Mendonça, M., Lacey, S., & Hvelplund, F. (2009). Stability, participation and transparency in renewable energy policy: Lessons from Denmark and the United States. *Policy and Society*, *27*(4), 379–398.

Metcalf, G. (2019). *Paying for Pollution: Why a Carbon Tax Is Good for America. Oxford University Press, USA.*

Nalule, V. R. (2021). How to Respond to Energy Transitions in Africa: Introducing the Energy Progression Dialogue. *Energy Transitions and the Future of the African Energy Sector* (pp. 3–35). Springer International Publishing.

Nasirov, S., O'Ryan, R., & Osorio, H. (2020). Decarbonisation tradeoffs: A dynamic general equilibrium modeling analysis for the Chilean power sector. *Sustainability (Switzerland)*, *12*(19).

Newell, P. (2020). The business of rapid transition. *WIREs Clim Change*, *e670*. Retrieved July 29, 2020, from https://doi.org/10.1002/wcc.670

OECD. (2014). *Cities and Climate Change: National Governments Enabling Local Action*.

Okpaleke, F., & Abraham-Dukuma, M. C. (2020). Dynamics of Resource Governance, Climate Change, and Security: Insights from Nigeria and Norway. *Journal of strategic security*, *13*(4), 123–140.

Olander, P. (2019). Economic Diversification and Institutional Quality—Issues of Concentrated Interests. *Studies in Comparative International Development*, *54*(3).

Onuigbo, R. A., & Innocent, E. O. (2015). State Governors and Revenue Allocation Formula in Nigeria : A Case of the Fourth Republic. *International Journal of Accounting Research*, *2*(7), 14–36. Al Manhal FZ, LLC.

OPEC. (2020). *2020 OPEC Annual Statistical Bulletin*. Retrieved from www.opec.org

Paul, F. C. (2018). Deep entanglements: History, space and (energy) struggle in the German Energiewende. *Geoforum*, *91*, 1–9.

Pearse, R. (2020). Theorising the Political Economy of Energy Transformations: Agency, Structure, Space, Process. *New Political Economy*.

PEPANZ. (2019). *An Introduction to New Zealand's Oil and Gas Industry*. Retrieved January 4, 2021, from https://www.pepanz.com/dmsdocument/16

Rosenbloom, D., & Rinscheid, A. (2020). Deliberate decline: An emerging frontier for the study and practice of decarbonisation. *WIREs Climate Change*, wcc669. Wiley-Blackwell.

Röttgers, D., & Anderson, B. (2018). Power struggle: Decarbonising the electricity sector - Effects of climate policies, policy misalignments and political economy factors on decarbonisation. *OECD Environment Working Papers, No. 139*.

Salon, D., Murphy, S., & Sciara, G.-C. (2014). Local climate action: motives, enabling factors and barriers. *Carbon Management*, *5*(1), 67–79.

Sovacool, B. K. (2016). How long will it take? Conceptualising the temporal dynamics of energy transitions. *Energy Research & Social Science*, *13*, 202–215. Retrieved July 19, 2020, from http://creativecommons.org/licenses/by/4.0/

Sovacool, B. K., Bergman, N., Hopkins, D., Jenkins, K. E. H., Hielscher, S., Goldthau, A., & Brossmann, B. (2020). Imagining sustainable energy and

mobility transitions: Valence, temporality, and radicalism in 38 visions of a low-carbon future. *Social Studies of Science, 50*(4), 642–679.
Stephenson, J., Barton, B., Carrington, G., Doering, A., Ford, R., Hopkins, D., Lawson, R., et al. (2015). The energy cultures framework: Exploring the role of norms, practices and material culture in shaping energy behaviour in New Zealand. *Energy Research and Social Science, 7.*
Stephenson, J., Barton, B., Carrington, G., Gnoth, D., Lawson, R., & Thorsnes, P. (2010). Energy cultures: A framework for understanding energy behaviours. *Energy Policy, 38*(10).
Szulecki, K. (2018). Conceptualising energy democracy. *Environmental Politics, 27*(1), 24–41.
Szulecki, K., & Overland, I. (2020). Energy democracy as a process, an outcome and a goal: A conceptual review. *Energy Research & Social Science, 69.*
The Abu Dhabi Government. (2020). *The Abu Dhabi Economic Vision 2030.*
The Kingdom of Saudi Arabia. (2020). *The Kingdom of Saudi Arabia Vision 2030.*
Tvedten, I., Lázaro, G., Jul-Larsen, E., & Agostinho, M. (2018). *Urban poverty in Luanda, Angola.* Massachusetts. Retrieved January 28, 2021, from www.cmi.no
UNDESA. (2019). *World Urbanisation Prospects: The 2018 Revision.* New York.
van Veelen, B., & van der Horst, D. (2018). What is energy democracy? Connecting social science energy research and political theory. *Energy Research & Social Science, 46*, 19–28.
Venture Taranaki. (2019). *Taranaki 2050 Roadmap: Our Just Transition to a Low-Emissions Economy.*
Wallach, O. (2021, June 8). Race to Net Zero: Carbon Neutral Goals by Country. *Visual Capitalist.* Retrieved October 3, 2021, from https://www.visualcapitalist.com/race-to-net-zero-carbon-neutral-goals-by-country/
Wesseling, J. H., Lechtenböhmer, S., Åhman, M., Nilsson, L. J., Worrell, E., & Coenen, L. (2017). The transition of energy intensive processing industries towards deep decarbonisation: Characteristics and implications for future research. *Renewable and Sustainable Energy Reviews, 79*, 1303–1313. Elsevier Ltd.
Wimbadi, R. W. (2020). From decarbonisation to low carbon development and transition: A systematic literature review of the conceptualisation of moving toward net-zero carbon dioxide emission (1995–2019). *Journal of cleaner production, 256*, 120307-.
World Bank. (2016, September 22). New Data-Driven Planning Tool Helps Cities Advance Climate Action. Retrieved October 3, 2021, from https://www.worldbank.org/en/news/feature/2016/09/22/new-data-driven-planning-tool-helps-cities-advance-climate-action
Zalengera, C., To, L. S., Sieff, R., Mohr, A., Eales, A., Cloke, J., Buckland, H., et al. (2020). Decentralisation: the key to accelerating access to distributed energy services in sub-Saharan Africa? *Journal of Environmental Studies and Sciences, 10*(3), 270–289.

CHAPTER 9

To Opt-in or to Cop Out: COP26 and the Policy Dynamics of Decarbonising African Cities

Ayodele Asekomeh, Obindah Gershon, and Smith I. Azubuike

Abstract The COP26 Glasgow Climate Pact appears to have kept alive the ambition of restricting temperature rises to 1.5 °C above pre-industrial levels. However, developing countries must translate the agreements into specific policies and change instruments in their home countries. Carbon

A. Asekomeh (✉)
Department of Accounting and Finance, Aberdeen Business School, Robert Gordon University, Aberdeen, UK
e-mail: a.asekomeh@rgu.ac.uk

O. Gershon
Centre for Economic Policy and Development Research (CEPDeR), Department of Economics and Development Studies, Covenant University, Ota, Nigeria

Eduardo Mondlane University, Maputo, Mozambique

S. I. Azubuike
Durham Law School, Durham University, Durham, UK

S. I. Azubuike et al. (eds.), *Decarbonisation Pathways for African Cities*, Palgrave Studies in Climate Resilient Societies,
https://doi.org/10.1007/978-3-031-14006-8_9

abatement agreements and the responsibility for financing climate change actions may be inimical to Africa's fragile economies which are often dependent on natural resources and carbon-emitting activities. The Advocacy Coalition Framework (ACF) helps to evaluate the policy subsystem to explain how coalitions' beliefs and resources can be channelled towards policymaking for the decarbonisation of African cities. Specifically, we use the ACF to review international cities coalitions and the Africa Adaptation Acceleration Program (AAAP) to explore the interactions and institutional settings needed to negotiate, agree and implement the Glasgow Climate Pact for decarbonising African cities.

Keywords Decarbonisation • African cities • COP26 Glasgow Climate Pact • Advocacy coalition framework • Policy • Governance • Stakeholders

9.1 Introduction

There has been renewed optimism that talks at the 26th Conference of Parties to the UN Framework Convention on Climate Change (UNFCCC), also known as COP26, have kept alive the ambition of restricting temperature rises to 1.5 °C above pre-industrial levels. Negotiations to complete the Paris Rulebook, originally proposed at COP21, continued for an extra day. The Paris Rulebook is intended to achieve a global agreement to accelerate climate action during the current decade (2020–2029). Its completion at COP26 is seen as real progress. However, the actual test of the outcomes is expected to arise from the follow-on action by delegates and Parties in their respective countries in translating the agreements to action (Obergassel et al. 2021). Specifically, delegates and governments of developing countries have their work cut out to operationalise the COP26 Glasgow Climate Pact (COP26 2021) as specific policy and change instruments in their home countries. This is because the most contentious issues at COP26 relate to the responsibility for the financing of climate change action and a lack of commitment to the kind of fossil fuel (particularly coal) abatement that would be required to maintain 1.5 °C.

We assess the Glasgow Climate Pact and the complexity of the choices before African cities in seeking to decarbonise, especially in determining how the agreements reached can be translated into changes in policies affecting cities. The nature of the talks and their implications may add to or distract from a decarbonisation agenda for African cities whereby the

Parties could either actively seek ways for opting into the measures or copping out from them given the burden that they may impose on their countries and economies. One of the contentious issues at COP26 was the shortfall in the funds that will be required to initiate and sustain a decarbonisation agenda for developing countries, and by extension, African cities. Developed countries have not fulfilled pledges agreed at the 21st Conference in Paris (COP21) to jointly provide mitigation and adaptation finance of USD100 billion annually by 2020 (Timperley 2021; Depledge et al. 2022). The pledge entailed offering relevant support through technology and capacity-building, which have also not been fully realised. This shortfall means that developing countries will struggle to implement climate change actions and they may not commit fully to or be capable of realising their nationally determined contributions (NDCs) to reduce emissions and manage climate change. Climate-vulnerable countries, especially given their dependence on natural resources whose extraction contributes to or worsens carbon emission, need developed countries to increase their level of climate financing (Timperley 2021).

When extrapolated to the development requirement of African cities, the funding constraints are further exacerbated by years of infrastructural and structural deficits and an unbroken trend of rural-urban migration (Mubangizi 2021; Selod and Shilpi 2021). African cities would require substantial new infrastructure financing, policy, and governance changes, and adoption of technology-related decarbonisation measures to help African countries meet NDCs. Cities in developed countries have structural advantages that are favourable or provide a basis for innovation and transformation (e.g., UK cities—Sait et al. 2018; Asekomeh et al. 2021). The approach to decarbonisation in African cities needs to be carefully framed to consider this important difference. Specifically, the systemic failings, structural and infrastructural gaps, and policy mismatch at the city level mean that climate adaptation and mitigation measures are needed, with the former previously often prioritised over the latter (Lwasa et al. 2018).

The dual pressures of limited funding and worsening infrastructural gaps mean that African cities are often struggling to break away from a vicious cycle that starts with improper planning and poor infrastructure funding and is reinforced by inadequate and insufficient access to grid and off-grid power sources and dysfunctional social structures that promote economic inequality and hinder social mobility and cohesion (Corfee-Morlot et al. 2019). Against this backdrop, developing countries must decide how they approach the subject of climate change action through seeking alternative funding arrangements, changes in policies, modification of their economic models, and adoption of new governance structures, with these measures being implemented from the city or settlement level. We employ elements of the Advocacy Coalition Framework (ACF) to analyse the different economic, financial, and governance challenges confronting developing countries that are the focus of policy making. We examine the changes that will be needed and issues that must be addressed if a shared view of the role of African cities in the attainment of climate objectives through decarbonisation is to be met through the coalescing of stakeholder advocacy efforts towards policy formulation.

Our use of the ACF involves an analysis of the so-called 'achievements' of the Glasgow Climate Pact, contextualised to the requirements for positioning African cities at the forefront of the decarbonisation agenda. The framework is used to explore coalitions involved in the policy landscape, especially for resource-rich developing countries where resource-based economies have created cities servicing the resource in question (e.g., Petro-cities like Port Harcourt and Luanda). The analysis considers the different coalition standpoints/beliefs that must be brokered in line with five ACF hypotheses if such countries are to opt in to the COP26 agreements. To this end, the framework offers insights for understanding how the peculiar attributes of coalitions in cities and their agendas can be coalesced into a common set of interests or policy positions to address carbon and emission challenges. Specifically, we use the ACF to review the policy subsystem to examine the sources of policy gaps due to coalitions' differing expectations for the role or place of African cities in the Glasgow Climate Pact. We also consider the institutional setting or options for negotiating, agreeing, and implementing measures for decarbonising African cities. We review funding, governance, economic and policy arrangements through the lenses of the Africa Adaptation Acceleration Program (AAAP) and international city alliances as examples of coalitions that would help the African cities' decarbonisation agenda. Based on this we highlight the specific resources these coalitions possess in furtherance of their policy-making agenda.

9.2 Contextual Review and Theoretical Framework

9.2.1 *The Decarbonisation Challenge*

The main takeaways from COP26 have been summarised as relating to increasing the drive for adaptation, mitigation, increased funding, and transparency in the disclosure of national actions (COP26 2021). From an African perspective, funding shortfalls remain a major constraint given the extent of the infrastructural deficit following decades of improper planning and disjointed approach to developing decarbonised cities. Adoption and implementation of adaptation measures (a more pertinent discussion than mitigation measures in this context) remains the focus. The perceived unfairness or unjustness of developing countries being at the receiving end of extreme climatic events attributable to climate change despite being minimal contributors to the problem adds to the funding conundrum (Okonjo-Iweala 2020). Figures 9.1 and 9.2 illustrate Africa's minuscule contribution to CO2 emissions from fossil fuel in absolute terms and per

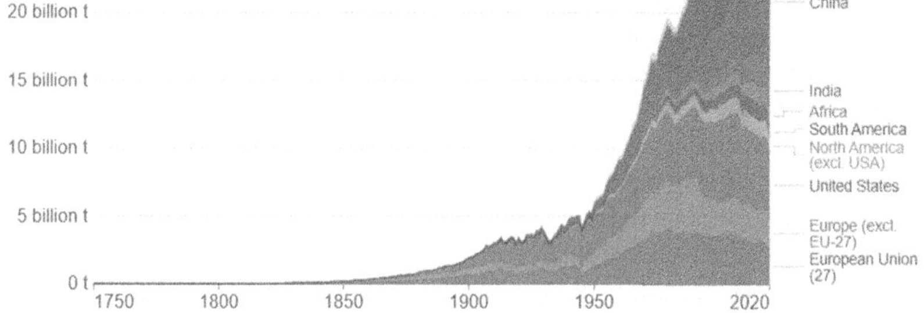

Fig. 9.1 Annual CO2 emissions from fossil fuels, by world region [Carbon dioxide (CO2) emissions from the burning of fossil fuels for energy and cement production. Land use change is not included] (Source: Global Carbon Project—Our World in Data [https://ourworldindata.org/co2-and-other-greenhouse-gas-emissions] CC BY)

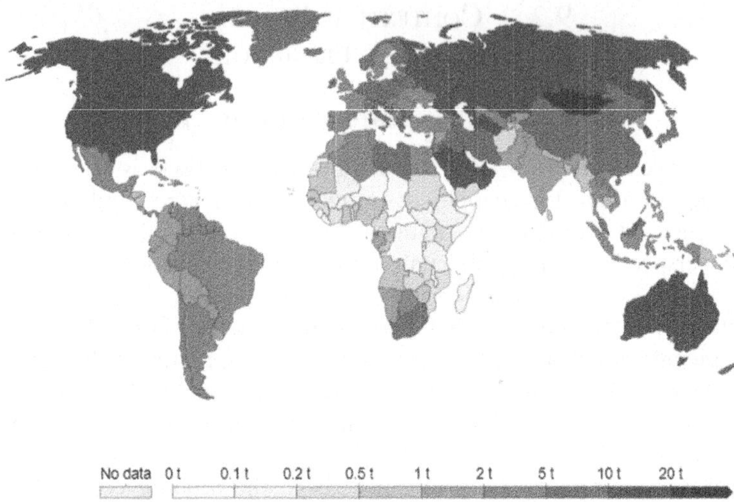

Fig. 9.2 Per capita CO2 emissions, 2020 [Carbon dioxide (CO2) emissions from the burning of fossil fuels for energy and cement production. Land use change is not included] (Source: Global Carbon Project—Our World in Data [https://ourworldindata.org/co2-and-other-greenhouse-gas-emissions] CC BY)

capita CO2 emissions, respectively. The implication of this is that the kind of fossil fuel (particularly coal) abatement that would be required to maintain 1.5 °C may not necessarily come from Africa, but the decarbonisation agenda for Africa through adaptation measures would mean that such countries can pursue a development and growth path that avoids the shortcomings of that followed by the developed world and increasingly by China and India, based on the burning of fossil fuels.

In this discourse, we explore decarbonisation gains to be had from focusing on the carbon emissions challenges of African cities. Support by developed countries will be crucial to helping developing countries to outline a clear strategy, particularly in relation to providing mitigation and adaptation finance, offering relevant support through technology and capacity-building, and through other indirect means (COP26 2021). This kind of support will make it possible to target specific aspects of need, particularly for cities where a host of coalitions have different expectations of what is required, but more importantly, the scale of intervention can be such that the burden and extent of the challenge can be broken down into smaller city-related needs. This will allow the use of intervention mechanisms which help to overcome "deep core beliefs" and "policy core beliefs" that are often rigidly held on to by coalitions thereby hindering the attainment of consensus for policy formulation (Sabatier and Weible 2007).

9.2.2 Translating the COP26 Glasgow Climate Pact to Policy for Cities

Four key *achievements* are now associated with COP26 (2021). First, the NDCs to reduce emissions from 153 countries and commitment to strengthening mitigation measures means that over 90% of world GDP now comes under the net zero commitments, with relevant guidelines and systems agreed. The finalised Paris Rulebook includes commitments to transition from coal power, accelerated move towards electric vehicles, the halting/reversing of deforestation, and reduction of methane emissions. Second, there is renewed commitment to dealing with climate impacts (i.e., minimising loss and damage) through adaptation action, with about 80 countries primed to address climate risks by either Adaptation Communications or National Adaptation.

Third, clear progress has been made towards the $100 billion climate finance goal by 2023 with 34 countries and five public institutions pledging to stop support for fossil fuels. It is envisaged that there will be a doubling of 2019 adaptation finance levels by 2025. The establishment of the Least Developed Countries Fund adds to arrangements by financial institutions and central banks to secure financing towards net zero. Fourth, COP26 has fostered collaboration between various stakeholders like governments, businesses, and civil society to rapidly deliver climate goals. The finalised Paris Rulebook provides a framework for transparent collaboration which could manifest in conversations about energy, electric vehicles, shipping, and commodities or in agreeing on standards for international carbon markets and common timeframes for emissions reduction targets.

Translating these preliminary agreements ('achievements') into specific action would require that multiple stakeholders or coalitions work towards the consensus that is needed to drive policy agendas at national levels. Ultimately, these should translate into specific measures for cities and settlements. The converse of this is also true, that is, the measures put in place at the city level will enhance the attainment of national contributions to carbon abatement targets at the national level. The ACF is a useful framework for explaining how such consensus can be brokered towards relevant policy formulation.

9.2.3 *The Advocacy Coalition Framework and Catalysing Action for Decarbonisation of African Cities*

The Advocacy Coalition Framework (ACF) proposed by Sabatier (1987), Sabatier and Jenkins-Smith (1999), and Sabatier and Weible (2007) offers different perspectives for understanding the policy process, perhaps only rivalled by the Institutional Analysis and Development Framework (IAD) (Sotirov and Memmler 2012). According to Cairney (2015, p.484), the ACF's eclecticism means that it is suitable for explicating complex policy-making systems characterised by decision-making under information constraints and high levels of uncertainty and ambiguity, especially where the time from decisions to outcomes could be up to a decade or more. In the case of climate change, net zero carbon emission targets are being agreed for mid-century, which is just about three decades away. Cairney also advocates ACF for systems involving multiple stakeholders and governmental levels and in which policy processing could range from highly politicised, publicly visible issues to specialist issues routinely handled by specialists outside the public domain. Adapting the Glasgow Climate Pact for African cities' decarbonisation fits this bill.

Figure 9.3 is a flow diagram depicting the main components of the ACF, which is typically used to hypothesise about the policymaking process. Public policymaking is construed as occurring within a 'policy subsystem' bounded by discernible and geographical attributes. We equate this to the national climate change policy subsystem where several advocacy coalitions could be competing for prioritisation of their views or beliefs to, for example, influence decisions by governmental authorities in relation to approaches to managing the climate change situation.

Based on Sabatier (1988), Sabatier and Jenkins-Smith (1993, 1999), and Sabatier and Weible (2007), coalitions may comprise disparate actors or stakeholders, with each coalition coalescing and coordinating actions around a common belief (the coalition "glue"—Cairney (2015, p.486)). Beliefs inform a coalition's actions and could be (1) deep core beliefs (underlying personal philosophy), (2) policy core beliefs (fundamental policy positions), or (3) secondary aspects (e.g., funding mechanisms, and specific information and institutions for delivering on and implementing policy goals). These three beliefs are of increasing changeability or

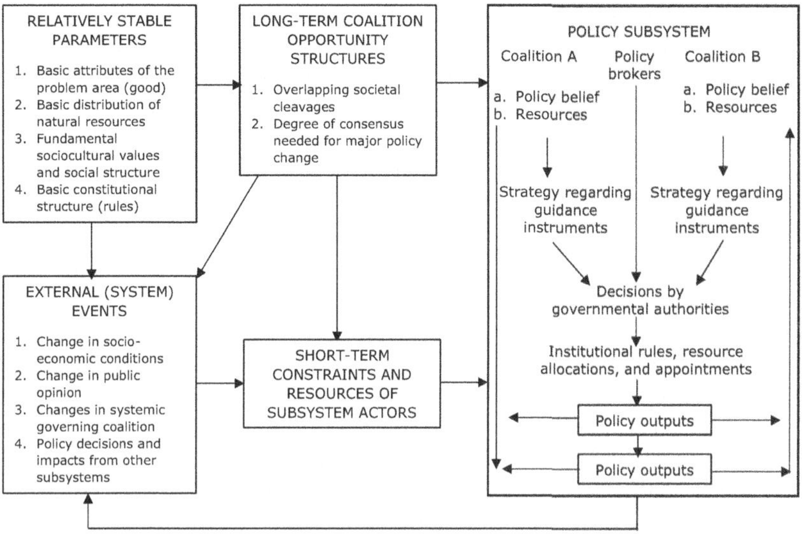

Fig. 9.3 The Advocacy Coalition Framework (Source: Sabatier and Weible 2007, p.202)

flexibility in the order of listing. Thus, normative and ontological deep core beliefs and policy core beliefs are the most difficult to change. In contrast, secondary aspects are more malleable (Sabatier and Jenkins-Smith 1999; Sabatier and Weible 2007). We explore the implications of this later.

Interactions of advocacy coalitions and facilitation of policy formulation (i.e., by policy brokers) in the policy subsystem take place within a wider system made up of three elements which impose short-term constraints on and determine the resources available to subsystem actors (Sabatier and Jenkins-Smith 1999; Sabatier and Weible 2007). The first element represents factors that are relatively stable over a decade or more like social values and constitutional structure. The second relates to long-term coalition opportunity structures which define the political system like the degree of consensus needed for policy change. The third consists of external (system) events, including changes in socio-economic conditions, government and public opinion, and policy decisions and impacts from other subsystems, which trigger behavioural responses in the policy subsystem.

The ACF is useful for explaining policy change deriving from changing the beliefs or views of the most influential coalitions. Major policy changes may require shifts in policy core values and minor changes, that is, in the secondary aspects of policy, may simply require modification of secondary aspects of beliefs (Sabatier and Jenkins-Smith 1999). Typically, deep core and policy core beliefs may change in response to extreme external shocks ("external perturbations") that alter the relative negotiating positions of coalitions while secondary aspect changes could arise through learning ("policy-oriented learning") involving the use of experience and new information to achieve enduring alternation of thought and behavioural intentions by revising existing or formulating new policies (Sabatier and Jenkins-Smith 1999, p.123; Sabatier and Weible 2007, p.198–199). Internal shocks (i.e., occurring within the policy subsystem) could also lead to a re-assessment and re-alignment of policy core beliefs and better intra- and inter-coalition understanding of the policy issues, which could result in "negotiated agreements" (Sabatier and Weible 2007, p.204–205). Policy changes from these consensual collaborative processes arise when coalitions are dissatisfied with the prevailing policies and cannot explore other avenues but are able to commit to independently mediated decision-making (Sabatier and Weible 2007, pp. 205–207).

9.2.4 Application of the Framework to the Glasgow Climate Pact

The ACF has been used severally in the environmental and energy policy space, especially in relation to developed countries or economies. We employ the ACF as a conceptual model to theorise about the relationships that are required to achieve a consensus of ideas, beliefs, and policy agendas to drive climate adaptation in African cities, where the policy subsystem relates to outlining institutional rules, resource allocations, and appointments within cities to promote adaptation and/or mitigation measures. We translate the Glasgow Climate Pact (COP26 2021) into considerations for the different coalitions likely to impact on the policy landscape for decarbonising African cities. Coalitions in this space could be varied, ranging from groups promoting the industry (usually that being serviced by the city) and the preservation of its place as a commerce or industrial centre and those looking to combat issues associated with modernisation of infrastructure and energy conservation, or countering the negative impacts of the industry's activities on the environment.

In this respect, the wider discourse in relation to climate change equates to external (system) events, with changes in public opinion about the impact of anthropogenic activities on the physical environment and climatic conditions. The need for sustainable socio-economic development in some of the world's poorest societies has seen Africa pursue changes in the systemic governing coalition through the African Union, the African Continental Free Trade Area (AfCFTA) agreement, and subtle arrangements like having a dedicated pavilion at COP26. Asekomeh et al. (2022) argue that a US or EU green deal type of intervention would be needed for a green post-COVID recovery in Africa. Dependence on natural resource rent and the expectation that such rent will continue to be accessible has informed the development of constitutional, regulatory and governmental structures (Mohamed 2020). These structures were previously mostly a top-down process, but are now entrenched in the democratic processes allowing grassroots (e.g., city council, representation). The central government usually manages and allocates the resource rent to units within the system, with the intention that these will be used, for instance, in city planning, transportation, waste management, and other subsystems. In some cases, the responsibilities could be categorised as centralised or devolved responsibilities, with city councils having control over specific city planning and development decisions (Natural Resource Governance Institute 2016; Asekomeh et al. 2021).

The intractability of the arrangements (often encapsulated in the constitution or legal framework) means that it represents a relatively stable parameter for policy making, along with fundamental socio-cultural values and social structures which create glaring divides between the affluent in urban neighbourhoods and the extremely poor slum dwellers within cities (Dang 2013). Finally, the protracted nature of the negotiations towards the Glasgow Climate Pact indicates that finding consensus for major policy change would remain a vital consideration for creating opportunities for long-term coalitions towards successful decarbonisation.

9.2.4.1 African City Coalitions and Competing Interests

Following Sotirov and Memmler (2012), we envision the African cities' policy subsystem as being made up of three advocacy coalitions. First, the "traditional city council management" paradigm would typically provide the basic city management framework, usually driven by the need to deliver city services guided by the value for money principle. Second, the "environmental and economic development oriented" coalitions typically challenge the traditional paradigm by seeking mainstream discussions about concerns for the environment and living conditions within cities, as well as the requirements for sustaining economic activities and opportunities for city dwellers. The third category is the "social concern" coalition, which is primarily interested in the place of cities in safeguarding the social well-being and livelihoods of city dwellers. For Petro-cities, these three advocacy coalitions often struggle to achieve consensus given that the central economic activity could directly be responsible for worsening both the environmental and social conditions. The role of policy brokers in helping cities adopt the COP26 Climate Pact will be in bringing these disparate coalitions to some form of compromise on pertinent policy matters.

The three coalitions identified above for cities represent groups that would need to adopt or be receptive to the Glasgow Climate Pact and would need to modify their policy core beliefs, at the very least, to make it possible for policy changes to be made to achieve the targets agreed. Accordingly, we turn our attention to how the policy subsystem can be construed and interpreted for the decarbonisation discourse in African cities. The ways the advocacy coalitions identified above would alter their beliefs or positions are usually framed as hypotheses by the ACF. We see the possibility of examining these relationships from five out of 15 commonly tested hypotheses listed by Weible et al. (2009, p.129):

Hypothesis 1: Actors within an advocacy coalition will show substantial consensus on issues pertaining to the policy core, although less so on secondary aspects.

Hypothesis 2: An actor (or coalition) will give up secondary aspects of the actor's belief system before acknowledging weaknesses in the policy core.

Hypothesis 3: Even when the accumulation of technical information does not change the views of the opposing coalition, it can have important impacts on policy—at least in the short run—by altering the views of policy brokers.

Hypothesis 4: Actors who share policy core beliefs are more likely to engage in short-term coordination if they view their opponents as (i) very powerful and (ii) very likely to impose substantial costs upon them if victorious.

Hypothesis 5: Actors who share (policy core) beliefs are more likely to engage in short-term coordination if they: (i) interact repeatedly; (ii) experience relatively low information costs; and (iii) believe that there are policies that, while not affecting each actor in similar ways, at least treat each fairly.

We provide discussions of how the COP26 Climate Pact provides insights for likely advocacy coalition interaction in relation to choices for African cities, especially in relation to issues which represent differences in beliefs or policy gaps.

9.2.5 *Coalition Differences or Gaps and Policy-making Constraints*

9.2.5.1 *Weak Institutional Capacities, Governance, and Regulatory Gaps*

One of the major challenges for policy making and implementation in relation to development of low-carbon cities and communities is weak institutional capacities. Although the Paris Agreement of 2015, the subsequent Rulebook agreed at COP26 and the United Nations Framework Convention on Climate Change (UNFCCC) aim to strengthen global responses to the threat of climate change, the differences in national capacities leave states in Africa with limited options for mainstreaming strategies to reduce greenhouse gas (GHG) emissions in the continent. To expedite action towards these responses, regulatory, policy, and institutional frameworks must set the stage to provide leadership and direction for a low-carbon economy. Governance and regulatory structures must be advanced above short-term political expediency and rent-seeking behaviour (Akinola 2018). Robust institutional and regulatory frameworks are required to effectively stimulate and facilitate the process of decarbonisation of African cities. Governance arrangements need strengthening, especially in promoting the participation of the private sector and promoting collaborative funding arrangements to build up carbon finance mechanisms and opportunities (Michaelowa et al. 2021).

While coalitions may have shared core beliefs on the need for these structures and capacities, the specific details or secondary aspects of the arrangements would be more contentious. This is in line with hypotheses 1 and 2. The need for climate action in cities is clearly a pursuit that the "traditional city council management", "environmental and economic development oriented", and "social concern" coalitions identified previously can rally around. However, the specific mechanisms for achieving this and responsibilities for key decisions would prove more difficult. For example, financing mechanisms may contravene value for money expectations for city council management even if the sustainability projects would be favourably received by environmental campaigners.

9.2.5.2 Planning Gap

For African cities, there is a vicious cycle of infrastructural deficit due to pressures created by rural-urban migration whereby the meagre development projects undertaken by city councils and authorities are immediately swamped by growth in city populations. This means that there is always a structural and infrastructural deficit when it comes to development. With the cities' carbon problem being exacerbated by migration from rural communities, it follows that the decarbonisation challenge in cities cannot be addressed in exclusion of the rural carbon and energy poverty challenge. The need to manage the influx of the rural community dwellers into cities pits different coalitions against one another. For instance, the "social concern" coalition may be interested in promoting inclusiveness and opportunities for upward social migration, but the burden on infrastructure like schools and housing may mean "traditional city council management" struggles to keep up. Breaking that economic vicious cycle through the ACF requires that parties or stakeholders make a commitment to changing their outlook and policy core beliefs.

In line with hypothesis 3, the data and technical information on distortionary impact of migration on city planning will be one that policy brokers need to consider even if the coalitions continue to hold different core beliefs.

9.2.5.3 Funding and Economic Gaps

A critical consideration, especially in cities, is the means and method of financing climate mitigation and resilience measures such as climate-smart systems for buildings and eco-friendly transport systems. Similarly, waste management and other green interventions are necessary for both decarbonisation and achieving the 2063 sustainable development agenda in Africa (African Union 2015). The financing challenge relates to incentivising investment in the decarbonisation of Africa's cities from the private sector (Michaelowa et al. 2021) given that public (city council) funding is more likely to be inadequate. This could be by means of grants and fiscal concessions leading to tax rebates for specific types of investments. These must reconcile explicitly with the economics and governance aspects of decarbonisation for climate resilience in Africa.

In the short run due to the potential stranding of assets, decarbonisation may put at risk Africa's economies that are overly dependent on natural resources (Ansari and Holz 2020). The challenge is to ensure that long-term economic and climate resilience is not ignored for short-term gains, with the knock-on effects on the infrastructural development of cities and communities. Since decarbonisation may mean less dependence on fossil-fuel sources and their revenue, it may be difficult for African resource-rich countries to secure resource-backed loans for development purposes (Landry 2018) and other forms of financing will be required.

The COP26 negotiations have shown that several actors hold the view that agreeing to carbon cuts will mean loss of resource rents. They opine that this will put the burden of climate mitigation on developing countries or countries whose stage of development means that fossil fuels are critical to their revenues and/or energy mix. This has seen Parties from these countries engaging in short-term coordination and working together to prevail on developed countries to commit to funding climate action (in line with hypotheses 4 and 5).

9.3 COP26 Opt-in Policy Formulation Mechanisms for African Cities

9.3.1 *Coalitions, Coalition Resources, and Policy Brokers for Decarbonisation of African Cities*

In addition to the discourses about the beliefs of coalitions that provide the basis for the formation of coalitions, the ACF also highlights the need for coalitions to possess and be able to deploy specific resources in support of their beliefs and towards influencing the policymaking process. We posit that to translate the COP26 Glasgow Climate Pact 'achievements' from the discussion table to specific policy instruments for cities in Africa, coalitions in the policy subsystem must possess the resources to engage with policy brokers in support of their policy core beliefs. These resources will determine whether they are able to address the previously identified policy gaps for decarbonising cities. To illustrate the deployment of these resources, in this concluding section, we examine a couple of coalitions or coalition structures that represent the possibility of deploying relevant resources towards promoting policy learning, and intra- and inter-coalition understanding of the policy issues and negotiation towards decarbonising African cities.

9.3.1.1 Alliance of Cities

Several coalitions and alliances based on cities have emerged to fill the gaps identified above. These include the United States' Clean Cities Coalition Network (https://cleancities.energy.gov/), the International Council for Local Environmental Initiatives—Local Governments for Sustainability (ICLEI) (https://www.iclei.org/), C40 Cities (https://www.c40.org/) and the Global Covenant of Mayors for Climate and Energy (https://www.globalcovenantofmayors.org/). Given their widespread membership (for instance, the C40 cities is a network of about 100 cities, including 13 African cities, collaborating on GHG emissions and currently make up more than 25% of the global economy with over 700 million people), these entities represent coalitions, and vitally, they also play an active role in facilitating policy brokering in recognition of the significance of cities in implementing climate action. They often provide mechanisms for producing and exemplifying innovative solutions and technologies for climate adaptation and mitigation. In this light, cities can be seen as providing the supporting mechanisms (institutions, private sector funding, etc.) that will foster innovation of mitigation and adaptation techniques that can be the basis for national action. These alliances clearly support the view of coalition interactions and coordination implied in hypotheses 4 and 5.

The benefits of membership of such alliances for African cities include providing a basis for designing and trialling innovative solutions in relation to transportation (e.g., mass transit networks), waste management, and water use efficiency. Innovations include efficient transport (including green buses and dedicated lanes) and traffic systems to alleviate traffic congestion and pollution from transport-related CO2 emissions, as well as provision of nudges by way of incentives for city dwellers to imbibe sustainable lifestyles (including reducing/recycling or reusing products) and use of city cleaning technologies. Membership provides a basis for knowledge and technology transfer to other cities and such cities could then set aspirational goals to provide benchmarks that exceed the COP26 targets for emissions abatement. Membership, involving representation by mayors and decision makers of network cities, also means that cities could provide a fulcrum for national climate initiatives starting with city initiatives. City representatives could also provide a governance basis by working with government agencies and parastatals at the national level and in international negotiations like the COP26 Climate Pact. This would ensure backward and forward feedback mechanisms for identifying and promoting climate mitigation and adaptation practices.

City alliances or networks offer further benefits in relation to bridging the finance and planning/infrastructural gaps identified previously. This is through mechanisms like providing access to climate-related infrastructure financing and equipping city representatives with the knowledge and skills for project finance budgeting and funds management. Finance facilities are accessible which offer competitive rates and flexibilities that would otherwise be difficult to secure, and which would otherwise mean that projects adding to climate resilience for cities are forfeited. In addition, infrastructural development and planning are often constrained by limited access to relevant data. Data from member cities offer a basis for comparison and benchmarking to measure progress and improve planning decisions. Requisite datasets could range from population and migration statistics to information about operational aspects like waste management that are relevant for building climate resilience.

Thus, coalitions of cities could facilitate the attainment of COP26 targets where such cities achieve an aspirational status for other cities in a country, ensuring that they all contribute to the NDCs for building national resilience and achieving carbon abatement in the national development agenda.

9.3.1.2 Africa Adaptation Acceleration Program

The Africa Adaptation Acceleration Program (AAAP), launched at the Climate Adaptation Summit in 2021, was jointly developed by the African Development Bank (ADB) and the Global Centre on Adaptation (GCA), with the latter specifically designated a global solutions broker, providing advocacy support for accelerating adaptation action. The AAAP operationalised the Africa Adaptation Initiative to raise about $25 billion in poverty alleviation, youth empowerment through entrepreneurship skills and job creation, and invest up to $7 billion towards climate-resilient infrastructure development working with/through "Multilateral Development Banks and other leading implementation organisations, stakeholders, and political and technical bodies" (AAAP 2021, p.2). These objectives are structured along four pillars which prioritise opportunities for climate adaptation and resilience.

The first pillar (Pillar 1) relates to the use of climate-smart digital technologies for agriculture and food security. In recognition of the dependence of most African economies on agriculture for food and employment, this pillar addresses the need to manage the sector's vulnerability to climate change by employing the right technologies to boost productivity. This will entail improving access (in terms of availability and affordability) and applicability of data-driven digital solutions to promote agricultural productivity, especially through the private sector. The second pillar (Pillar 2) is the Africa infrastructure resilience accelerator, which is intended to bridge the infrastructure deficit of about $130 billion–$170 billion a year, with an estimated additional investment of only 3% to total costs to make such infrastructure resilient and capable of delivering on a significant number of the sustainable development goals (SDGs), the Paris Agreement (and by extension the COP26 Climate Pact) and the Sendai Framework (AAAP 2021).

Pillar 3 relates to empowering youth for entrepreneurship and job creation in climate adaptation and resilience, to leverage Africa's young population (expected to double to over 830 million by 2050) through creation of economic activities, and "drive resilience through their innovativeness, energy, and entrepreneurship" (AAAP 2021, p.5). The fourth pillar (Pillar 4) provides the crucial backing for achieving the objectives of the AAAP through innovative finance initiatives. Global climate finance received by Africa is only a microcosm (4%) of the average annual finance of $30 billion per annum as of 2017–2018, with an even smaller fraction devoted to adaptation and resilience initiatives and funding has been severely curtailed by the global response to the COVID-19 pandemic.

Collectively, these pillars are primed to help African countries and, by extension, cities achieve decarbonisation as they address the specific gaps identified previously. They therefore represent policy brokers or coalition enablers within the policy subsystem in the ACF. The AAAP thus provides a basis for coalitions' formation and coordination through the policy core beliefs outlined, in line with hypotheses 4 and 5. The AAAP provides rich development-related data that will inform the decisions/choices of policy brokers in line with hypothesis 3.

9.3.2 *Discernible Coalition Resources*

From the example of the coalitions identified above, clear coalition resources itemised by Sabatier and Weible (2007, p.201–202) can be articulated based on the ACF. The city alliances and the AAAP are backed by relevant agreements that confer (a) *formal legal authority to make policy decisions*. For instance, mayors of cities bring needed credibility, financial clout from their budgets, and potential for international networking. The city alliances and AAAP have also been useful in shaping (b) *public opinion* on matters of climate change and decarbonisation. They are also sources of veritable (c) *information* regarding the severity and the urgency of the problem. Their standing also means that they are capable of commissioning research or studies that offer credible and free-of-bias perspectives to support policy arguments based on their beliefs.

In addition, city alliances and the AAAP are resourced by (d) *mobilizable troops* to promote the beliefs and activities of these coalitions and to bring their policy arguments to the consciousness of the wider public. The vast pooling of (e) *financial resources* and access to funds provided by membership of city alliances and by instruments created by the AAAP means that the capability for organising other resources, conducting research, producing information, and shaping public opinion has increased several folds. Most importantly, city alliances and AAAP provide (f) *skilful leadership* or offer a good basis for building such leaders and shaping thought leadership regarding the beliefs of the coalition, which would go a long way to influencing policy making.

Operationalising the Glasgow Climate Pact would need the dominant coalitions to use these resources, guided by policy brokers, towards securing the best outcomes for African cities, striking a balance between achieving decarbonisation objectives and doing so in ways that do not severely impact on the economic and social situations of cities. This will be the basis on which the Climate Pact can be truly judged to be just and equitable for the parties involved.

9.4 Summary and Conclusion

The ACF approach provides a perspective for understanding how various coalitions or stakeholders can make choices in African cities and communities to align with global climate action. The messages from COP26 would need to be communicated to relevant coalitions to attempt to achieve consensus of beliefs (from policy core beliefs to secondary beliefs or specific aspects) and approaches to assist Africa in creating decarbonised cities by limiting carbon-emitting sources and eliminating practices that contribute to GHG emissions within cities. This goal should be pursued alongside providing mechanisms for much-needed development to help city dwellers escape the poverty trap that is birthing several slums around resource-exploiting cities. The ACF approach to adopting the COP26 agreements and decarbonisation intervention as described here align with the Sustainable Development Goals' (SDGs) pursuit of affordable and clean energy (SDG 7) and the development of sustainable cities and communities (SDG 11).

Buildings, waste disposal systems, and commercial activities within cities are not carbon neutral. Some of these practices or activities are ignored in Africa's low-carbon and climate action discourse. A well-thought-out strategy and policy direction are crucial in achieving the much-needed decarbonisation in Africa's settlements. It is crucial to be able to catalyse action by getting disparate coalitions or groups to coalesce in support of the requisite policy instruments and choices that need to be made. We identified specific elements of the ACF that would provide African cities the basis for building or leveraging on coalitions to achieve consensus that are required to translate the Glasgow Climate Pact into specific policies to achieve decarbonisation. Policymaking for decarbonising African cities through the ACF involves understanding coalitions' beliefs and resources at their disposal. It explicates how beliefs and resources can be combined to reinforce the process through policy-oriented learning and negotiations, possibly in response to external shocks or perturbations.

References

AAAP (2021). Africa Adaptation Acceleration Program, Global Center on Adaptation, https://www.afdb.org/sites/default/files/2021/10/19/africa_adaptation_acceleration_program.pdf.

African Union. (2015). Agenda 2063: The Africa we want. African Union Commission, September 2015. Retrieved from https://au.int/sites/default/files/documents/36204-doc-agenda2063_popular_version_en.pdf.

Akinola, A.O., 2018. Africa's Quest for Industrialization and Development: Rethinking Rents and Rent-Seeking. *African Renaissance*, 15(4), pp.9-28.

Ansari, D. and Holz, F. (2020). Between stranded assets and green transformation: Fossil-fuel-producing developing countries towards 2055. World Development, 130, p.104947, https://doi.org/10.1016/j.worlddev.2020.104947.

Asekomeh, A., Gershon, O. and Azubuike, S.I. (2021). Optimally Clocking the Low Carbon Energy Mile to Achieve the Sustainable Development Goals: Evidence from Dundee's Electric Vehicle Strategy. *Energies*, 14, 842. https://doi.org/10.3390/en14040842

Asekomeh, A., Azubuike, S.I. and Gershon, O. (2022). Post-COVID-19 and African Agenda for a Green Recovery: Lessons from the European Union and the United States of America. In: E. Osabuohien et al., eds. COVID-19 in the African Continent: Sustainable Development and Socioeconomic Shocks. Emerald Publishing (ISBN 978-1-80117-687-3). https://doi.org/10.1108/978-1-80117-686-620221028

Cairney, P. (2015). Paul A. Sabatier, "An advocacy coalition framework of policy change and the role of policy-oriented learning therein". In: Balla, S. J., Lodge, M. and Page, E.C. (Eds), The Oxford handbook of classics in public policy and administration, Oxford University Press: Oxford, 484–497.

Corfee-Morlot, J., Parks, P., Ogunleye, J., and Ayeni, F. (2019). Achieving Clean Energy Access in Sub-Saharan Africa, Retrieved from https://www.oecd.org/environment/cc/climate-futures/case-study-achieving-clean-energy-access-in-sub-saharan-africa.pdf

COP26. (2021) Presidency Outcomes – COP26: The Glasgow Climate Pact. https://ukcop26.org/wp-content/uploads/2021/11/COP26-Presidency-Outcomes-The-Climate-Pact.pdf.

Dang, D. Y. (2013). Revenue allocation and economic development in Nigeria: An empirical study. *SAGE Open*, *3*(3), https://doi.org/10.1177/2158244013505602

Depledge, J., Saldivia, M. and Peñasco, C. (2022). Glass half full or glass half empty? The 2021 Glasgow Climate Conference. *Climate Policy*, 22(2), pp.147-157, https://doi.org/10.1080/14693062.2022.2038482.

Landry, D. (2018). The risks and rewards of resource-for-infrastructure deals: Lessons from the Congo's Sicomines agreement. *Resources Policy*, 58, pp.165-174, https://doi.org/10.1016/j.resourpol.2018.04.014.

Lwasa, S., Buyana, K., Kasaija, P. and Mutyaba, J. (2018). Scenarios for adaptation and mitigation in urban Africa under 1.5 C global warming. *Current opinion in environmental sustainability*, 30, 52-58, https://doi.org/10.1016/j.cosust.2018.02.012.

Michaelowa, A., Hoch, S., Weber, A-K., Kassaye, R. and Hailu, T. (2021) Mobilising private climate finance for sustainable energy access and climate change mitigation in Sub-Saharan Africa, *Climate Policy*, 21:1, 47-62, https://doi.org/10.1080/14693062.2020.1796568.

Mohamed, E. S. E. (2020). Resource rents, human development and economic growth in Sudan. *Economies*, *8*(4), 99. https://doi.org/10.3390/economies8040099

Mubangizi, B.C. (2021). Rural-Urban Migration and Smart Cities: Implications for Service Delivery in South Africa. *African Renaissance*, 18(1), (1744-2532).

Natural Resource Governance Institute. (2016, September). Natural Resource Revenue Sharing. https://resourcegovernance.org/sites/default/files/documents/nrgi_undp_resource-sharing_web_0.pdf

Obergassel, W., Arens, C., Beuermann, C., Brandemann, V., Hermwille, L., Kreibich, N., E. Ott, H., & Spitzner, M. (2021). Turning Point Glasgow? An Assessment of the Climate Conference COP26. *Carbon & Climate Law Review*, Volume 15, Issue 4, pp. 271–281. https://doi.org/10.21552/cclr/2021/4/4

Okonjo-Iweala, N. (2020). Africa can play a leading role in the fight against climate change. https://www.brookings.edu/research/africa-can-play-a-leading-role-in-the-fight-against-climate-change/ (accessed 14 March 2022).

Sabatier, P.A. (1987). Knowledge, policy-oriented learning, and policy change. An advocacy coalition framework. *Knowledge: Creation, Diffusion, Utilization*, 8 (4), 649–692.

Sabatier, P.A. (1988). An advocacy coalition framework of policy change and the role of policy-oriented learning therein. *Policy Sciences*, 21, 129–168.

Sabatier, P.A. and Jenkins-Smith, H.C. (1993). The Advocacy Coalition Framework: assessment, revisions, and implications for scholars and practitioners. In: Sabatier, P.A., Jenkins-Smith, H.C. (Eds.), *Policy Change and Learning: An Advocacy Coalition Approach*, Westview Press: Boulder (Co), 211–235.

Sabatier, P.A. and Jenkins-Smith, H.C. (1999). The Advocacy Coalition Framework: an assessment. In: Sabatier, P.A. (Ed.), *Theories of the Policy Process.* Westview Press: Boulder (Co.), 117–166.

Sabatier, P.A. and Weible, C.M. (2007). The Advocacy Coalition Framework: Innovations and Clarifications. In: Sabatier, P. (Ed.), *Theories of the Policy Process.* Westview Press: Boulder (Co.), 189–220.

Sait, M.A., Chigbu, U.E., Hamiduddin, I. and De Vries, W.T. (2018). Renewable energy as an underutilised resource in cities: Germany's 'Energiewende' and lessons for post-Brexit cities in the United Kingdom. *Resources*, 8(1), 7; https://doi.org/10.3390/resources8010007.

Selod, H. and Shilpi, F. (2021). Rural-urban migration in developing countries: Lessons from the literature. *Regional Science and Urban Economics*, 91, p.103713, https://doi.org/10.1016/j.regsciurbeco.2021.103713.

Sotirov, M. and Memmler, M. (2012). The Advocacy Coalition Framework in natural resource policy studies — Recent experiences and further prospects, Forest Policy and Economics, 16, 51-64, ISSN 1389-9341, https://doi.org/10.1016/j.forpol.2011.06.007.

Timperley, J. (2021). The broken $100-billion promise of climate finance – And how to fix it. *Nature*, 598(7881), pp.400-402, https://doi.org/10.1038/d41586-021-02846-3.

Weible, C.M., Sabatier, P.A. and McQueen, K. (2009). Themes and variations: Taking stock of the advocacy coalition framework. *Policy studies journal*, 37(1), pp.121-140, https://doi.org/10.1111/j.1541-0072.2008.00299.x.

CHAPTER 10

Conclusion: Towards a Decarbonisation Framework for African Cities

Obindah Gershon, Smith I. Azubuike, and Ayodele Asekomeh

Abstract The drive to decarbonise African cities is significantly disruptive due to their income level and dependence on carbon-emitting energy sources. As such, the policy and strategic governance of decarbonisation pathways need to be pursued within this perspective towards optimally

O. Gershon (✉)
Centre for Economic Policy and Development Research (CEPDeR), and Department of Economics and Development Studies, Covenant University, Ota, Nigeria

Eduardo Mondlane University, Maputo, Mozambique
e-mail: obindah.gershon@covenantuniversity.edu.ng

S. I. Azubuike
Durham Law School, Durham University, Durham, UK

A. Asekomeh
Department of Accounting and Finance, Aberdeen Business School, Robert Gordon University, Aberdeen, UK

S. I. Azubuike et al. (eds.), *Decarbonisation Pathways for African Cities*, Palgrave Studies in Climate Resilient Societies,
https://doi.org/10.1007/978-3-031-14006-8_10

contributing to climate change reduction. The chapter summarily posits that proposed/adopted low-carbon options require integrated plans for their actualisation. Moreover, such integrated plans should consider the interconnection between carbon emission and economic sustenance on the one hand. On the other hand, the need exists for indigenous technological innovation to sustain the pursuit of decarbonisation in African cities. It concludes by asserting that decarbonising African cities presents opportunities for reducing the burden of diseases and other environmental challenges in Africa.

Keywords Decarbonisation framework • African cities • Economic sustainability • Governance strategies

10.1 Introduction

The drive to decarbonise African cities involves shifting from dependence on carbon-based energy resources to renewable energy resources. This shift is necessary primarily as the region depends on carbon-emitting sources to power machines, provide heating and lighting for homes, cook food, generate electricity, and for economic sustainability. As such, the decarbonisation process requires novel ways of doing things, thinking, and innovative technologies. However, there will be disruptions, and managing these disruptions in the decarbonisation process will not be easy (Asekomeh et al. 2022). This is because most activities and practices in Africa involve some form of carbon emissions (Gershon and Asaolu 2020; Egbetokun et al. 2020). In this regard, we posit that the decarbonisation agenda must be pursued in ways that are peculiar to African societies to optimally contribute to achieving global climate action (Asekomeh et al. 2022). It means that the low-carbon process ought to consider the current challenges and how to address them using renewable energy strategies and low-carbon policies while also allowing the various economies and industries to thrive.

Given the overall theme of this book, the authors focused mainly on low-carbon strategies and governance and policy approaches for African cities in the decarbonisation process. This dimension is indispensable as the necessary foundation for low-carbon cities must be laid to ensure the operationalisation of low-carbon cities in Africa. Therefore, various authors in this book have highlighted several approaches and governance directions towards the decarbonisation of African cities.

10.2 Approaches Towards Cities' Decarbonisation in Africa

African cities need to integrate solar urban planning into city design to enhance decarbonisation. Akrofi, Okitasari, Ohunakin, and Azubuike discussed this in **Chap. 2** by advancing the challenges and prospects of achieving decarbonisation on the foundations of solar urban planning in Africa. The Chapter notes that advances in renewable energy applications, especially solar photovoltaics (PV), provide an enormous opportunity to advance decarbonisation and clean energy access in African cities. Thus, an effective strategy for achieving this goal is through the integration of solar PVs on buildings. According to the International Energy Agency (IEA 2014), this strategy can utilise more than half of the global solar capacity by 2050. This integration is crucial in cities, which currently consume about 78% of the energy produced globally, and account for more than 60% of greenhouse gas emissions. However, some of the challenges in achieving this strategy can be attributed to the failure to consider Building Integrated Photovoltaics (BIPV) at the early stages of the urban design/planning process. The non-integration of this strategy often leads to suboptimal and unattractive outcomes, which fail to stimulate the uptake of these technologies in cities. To overcome this challenge, Akrofi, Okitasari, Ohunakin, and Azubuike advance that solar energy concerns need to be integrated early in the urban planning process, hence the concept of solar urban planning. Implementing solar energy in urban planning deals with the connections between solar energy and urban morphology, land use, the spatial structure of cities, regulations, and socio-demographic factors.

Similarly, waste management operations can be designed to consider low-carbon practices instead of incineration. The scale of human waste, agricultural waste, animal waste, air, land, and sea pollution around cities, and carbon emissions arising from incineration present real challenges for climate action. Interestingly, the job-creation prospects and macroeconomic potentials of recycling waste in many African cities are yet to be harnessed. So, in **Chap. 3**, Muniafu and Nzembi propose contextualising waste management operations in Africa, asserting that governments ought to prioritise waste management. The issues are examined in major African cities (Cape Town, Dar-es Salaam, Koshe, Lagos, Nairobi, and Ouagadougou) with linkages to health (SDG 3) and access to clean water and sanitation (SDG 6). Furthermore, they highlight the possibilities of decarbonising African cities by focusing on recycling through a public-private partnership. The authors posit that strong legislations should be enacted and implemented with good standards at the national and municipal levels.

Urban healthcare facilities in Africa could apply the Top-Runner approach and climate-smart strategies to reduce their carbon footprint. The health sector is a heavy carbon emitter due to the products and equipment used in health facilities and the supply chain process for medical materials. For health facilities in Africa, the poor state of rural health facilities leaves the facilities in urban centres overburdened. Ironically, the healthcare sector that cares for patients—victims of poor waste management—also generates medical wastes that are hazardous to humans and the environment. Azubuike and Adeyemi, in **Chap. 4**, consider how innovative strategies such as the Top Runner approach and climate-smart strategies can help the decarbonisation process in healthcare facilities in cities. They frame their discussion on ecological modernisation theory and highlight the need for renewable energy and energy-efficient products, including practices like transport share to decarbonise cities.

The telecommunications sector must move from a fossil fuel-based energy system to a carbon-neutral energy system in their cell sites (Gershon and Agbene, 2021). In **Chap. 5**, Okundamiya and Wara explore decarbonisation pathways for the telecommunication sector. Africa's connection to the world via modern information and communications technology (ICT) is growing rapidly. This growth means more cell sites for telecommunication connections. As such, new cell sites require a carbon-neutral energy system to power these cell sites instead of the fossil fuel-fired energy system. The authors utilised a case study with 22-year meteorological datasets to propose that telecommunications firms switch from fossil fuel-fired to carbon-neutral energy systems in their cell sites. They employ the energy-equilibrium procedures to optimise the model and show that the wind-photovoltaic system is the optimal hybrid option for decarbonising telecommunication activities.

Grid-linked photovoltaic systems and fuel cell microgrid systems offer the possibility of reducing GHG emissions. As such, in **Chap. 6**, Wara and Okundamiya analyse the low carbon choices for a University campus within Nigeria for sustainably generating energy. They show that a combination of grid-linked photovoltaic system and fuel cell microgrid system offers the possibility of reducing GHG emissions from most university campuses in cities within Nigeria. Notably, most public and private Universities in Nigerian cities use a combination of public power supply (intermittent in supply) and minigrid-diesel generator system characterised by high and regular GHG emissions. Therefore, adding a solar-powered system will enhance carbon reduction, especially for universities that run primarily on diesel.

Governments in Africa can play a role in the decarbonisation drive. Muniafu, Ombewa, and Nzembi, in **Chap. 7**, focus on efforts by the Kenyan government, private sector, and civil society organisations to reduce GHG emissions. The authors capture different climate change mitigation actions that cut across cities—including strides in energy diversification to reduce dependence on fossil fuels. The seventh chapter also highlights a unique issue inherent in reducing carbon emissions within African urban cities—sustaining economic growth with a rapidly increasing population while addressing energy poverty amidst worsening income. Amongst the unique proposition in the chapter is the need for Kenyan (and other African) cities to develop strategies for recycling clean energy technology waste—like solar panels. This reinforces the discourse in Chap. 2 and proposes a job-creating pathway for addressing the environmental consequences of mass production and increasing the utilisation of clean energy appliances.

Political-economic considerations and governance reconfigurations for the future are at the core of effective carbon reduction strategies and climate action implementation. In **Chap. 8**, Abraham-Dukuma, Aholu, Nyokabi, and Dioha used theoretical and content analysis on four major Petro-cities (Hassi Messaoud, Luanda, Port Harcourt, and Tripoli). The Chapter discusses the governance, political, economic, and institutional interplays relevant to the decarbonisation processes of Petro-cities. From the analysis, the authors identify options for carbon emissions reduction, viz: optimising energy culture, promoting municipal energy democracy, and energy decentralisation. They assert that such regulatory, policy and institutional governance frameworks set the stage for a low-carbon economy. Alongside the policy actions for decarbonisation, a strategic commitment to institution-building is another essential intervention needed for sustainably mitigating climate change. This dimension is critical to the decarbonisation agenda to avoid inconsistent and inadequate regulations and institutions in Africa.

Africa requires new infrastructure to meet nationally determined contributions (NDCs). Asekomeh, Gershon, and Azubuike note in **Chap. 9** that despite the hope that COP26 offers in mitigating climate change, African delegations at COP26 face the challenge of translating some of the agreements to specific policy and change instruments in their home countries. This challenge is made more daunting because African cities require substantial new infrastructure to meet nationally determined contributions (NDCs). Moreover, developing countries would need to see more commitment from developed countries towards fulfilling pledges (of COP21) on financing, capacity building, and technology transfer. Accordingly, Asekomeh, Gershon, and Azubuike employ the Advocacy Coalition Framework (ACF) to analyse how the beliefs and resources available to various coalitions within the policy subsystem can be used to operationalise the 'achievements' of the Glasgow Climate Pact as specific decarbonisation measures for African cities. The coordination observed in various alliances of cities and the Africa Adaptation Acceleration Program (AAAP) indicate that while external perturbations (like the climate emergency) would mean that coalitions are willing to reconsider their policy core beliefs, policy-oriented learning and negotiations to support policy change by considering secondary beliefs of coalitions will be required to agree to specific policy instruments to achieve decarbonisation.

10.3 Looking Ahead

An examination of the chapters in this book reveals that the authors agree that global climate change needs to be addressed and that Africa's involvement in the drive for a low-carbon society is indispensable. Although the focus of the authors relates to various sectors of the economy in Africa, they all advance logical and coherent approaches towards the process of decarbonisation in African cities. It is evident that carbon-neutral practices in African cities will enhance the transition away from carbon-intensive activities. It is also essential for African economies to mainstream governance and policy directions to support the decarbonisation process. Albeit low-carbon practices and approaches require an integrated plan to actualise. However, this integrated plan should consider the interconnection between carbon emission and economic sustenance, as well as the need for technology to support the transition process. The reason is that the economic capacities of various countries vary. As such, the transition must consider this factor to achieve a just transition to low-carbon cities in

Africa. While these considerations are necessary, decarbonising African cities presents excellent opportunities for the continent as it reduces the burden of disease and other negative impacts of climate change in the continent of Africa. This latter objective drives the pursuit of decarbonisation in African cities.

References

Asekomeh, A., Gershon, O., & Azubuike, S. I. (2021). Optimally clocking the low carbon energy mile to achieve the Sustainable Development Goals: Evidence from Dundee's Electric Vehicle Strategy. *Energies, 14*, 842. https://doi.org/10.3390/en14040842

Asekomeh, A., Azubuike, S. I., & Gershon, O. (2022). Post-COVID-19 and African Agenda for a Green Recovery: Lessons from the European Union and the United States of America, in Osabuohien, E., Odularu, G., Ufua, D. and Osabohien, R. (Ed) (2022) COVID-19 in the African Continent (pp. 309-322). *Emerald Publishing Limited, Bingley*. https://doi.org/10.1108/978-1-80117-686-620221028

Egbetokun, S., Osabuohien, E., Onanuga, O., Akinbobola, T., Gershon, O., & Okafor, V. (2020). Environmental Pollution, Economic Growth and Institutional Quality: Exploring the Nexus in Nigeria. Management of Environmental *Quality,31*(1), 18-31. https://doi.org/10.1108/MEQ-02-2019-0050.

Gershon, O., & Asaolu, K. (2020). Evaporative quality of Nigeria's gasoline: truck loading perspective. Energ. Ecol. and Environ. https://doi.org/10.1007/s40974-020-00184-0

Gershon, O., & Agbene, E. (2021). Adopting Hybrid Energy Technology for Carbon Emissions Reduction in Nigeria's Telecommunications Industry. *IOP Conf. Ser.: Earth Environ. Sci. 655,* 012048, https://doi.org/10.1088/1755-1315/655/1/012048

IEA. (2014). Technology Roadmap Solar Photovoltaic Energy - 2014 edition. www.iea.org

Isaac, J. O., Olurinola, I. O., Gershon, O., & Aderounmu, B. (2021). Working Conditions and Career Aspirations of Waste Pickers in Lagos State. *Recycling, 6*(1), 1; https://doi.org/10.3390/recycling6010001

Index

A

Adaptation communications, 163

Advocacy Coalition Framework (ACF), 160, 163–166, 168, 170, 171, 174–176

Africa, 38–47

Africa Adaptation Acceleration Program (AAAP), 160, 173–175

African cities, 158–176

African petro-cities, 136–152

Algeria, 137, 139, 140

Alliances

- city, 160, 173, 175
- networks, 173

Angola, 137, 139, 140

Architectural design, 18

B

Beliefs

- deep core, 163–166
- policy core, 163–166, 168, 170, 171, 174, 175
- secondary aspects, 164–166, 168, 169

Buildings, 16, 22, 25, 27, 28

Built environment, 29

C

Capacity-building, 159, 162

Carbon abatement agreements, 163, 173

Carbon emissions, 52–61, 64, 67

Cedarcrest Hospital, 53, 57, 59, 65, 67

Cellular generation sites, 75

Cellular mobile technology, 76

Cities, 16–33

Clean Development Mechanisms (CDMs), 45, 46

Climate change, 52, 54–56, 58, 60, 62

Climate change mitigation, 125–126, 128

Climate financing, 159

Climate-vulnerable countries, 159

S. I. Azubuike et al. (eds.), *Decarbonisation Pathways for African Cities*, Palgrave Studies in Climate Resilient Societies,
https://doi.org/10.1007/978-3-031-14006-8

CO_2 emissions, 161, 162, 172
Coalition
coalition beliefs, 164
coalition resources, 171–175
environmental and economic development oriented coalition, 167, 169
International cities coalitions, 160
social concern coalition, 168–170
traditional city council management coalition, 170
Cohesion, 159
Commitments, 158, 163, 170
Consensus, 163–168, 175, 176
Consumption-based petro-cities, 151
COP26
agreements, 160, 176
COP26 Glasgow Climate Pact, 158, 163–164, 166–169, 171, 175, 176
COP26 targets, 172, 173
Cost of electricity, 74
Cost of energy (COE), 83, 87–90

D
Decarbonisation, 16, 32, 158–166, 168–176
Decarbonised cities, 144
Diesel generator (DG), 77, 83, 85, 87, 88, 97, 102–103, 105, 107, 110
Distributed generation, 29

E
Ecological modernisation (EM), 53–57, 67
Economic, 159, 160, 167–171
Electric grid design, 99–100
Electricity, 16–18, 24, 53, 63, 67
Electricity demand, 74
Electric power grid, 99
Electrolyser technology, 100
Emission reduction, 110
Energy, 52–59, 61–67
culture, 145–146, 151
decentralisation, 145, 147–148, 151
democracy, 145, 147–148, 151
independence, 26, 29
storage system, 80–81
transition, 138–141, 144, 147, 149, 151
Environment, 53–56, 58–60, 62, 66, 67
Expert engagement, 145

F
Financing climate change actions, 158
Fossil, 158, 161–163, 171
Fossil fuel, 52, 53, 58, 63, 64, 67, 74, 90
Frameworks
institutional and regulatory frameworks, 169
policy and institutional frameworks, 169
Fuel, 158, 161–163, 171
Fuel cell (FC) system, 97, 98, 100–102
Fuel combustion, 102
Funding, 159–161, 164, 169–172, 174
Future scenarios, 142–144, 151

G
Gap
funding and economic gaps, 170–171
institutional capacities, 169–171
planning gap, 170
policy gap, 160, 169, 171
regulatory gap, 169–171

Governance, 18, 136–152, 160, 169–171
Greenhouse gas (GHG), 52
Grid, 96, 98–100, 105, 107, 109, 110

H
Hassi Messaoud, 137, 140
Healthcare decarbonisation, 54–57
Health facilities, 52, 53, 57, 63, 65, 67
Hospital, 53, 57, 58, 61–67
Housing, 27
Hybrid energy system, 85, 87
Hybrid Optimisation Model for Electric Renewables (HOMER), 78, 81, 82, 90
Hybrid power system, 74–90, 97–99, 106–111
Hydrogen (H_2), 97, 100, 102
Hydrogen tank, 99, 100

I
ICT infrastructure, 105, 106, 108–110
Inequality, 159
Inflation index, 88
Inflation rate, 82, 88, 89
Information and communication technology (ICT), 99, 104
Infrastructural and structural deficits, 159
Internal rate of return (IRR), 108

K
Kenya, 118–132

L
Lagging petro-cities, 144
Land use, 17, 25, 27–28
Libya, 141
Local innovation, 131
Low-carbon, 53–57, 61, 62, 64, 65
Luanda, 137, 140

M
Market-deficit petro-cities, 142–143
Medical waste, 58, 65
Microgrid system, 97, 98
Mitigation and adaptation finance, 159, 162
Mobile telecommunication, 76

N
National adaption, 163
Nationally determined contributions (NDCs), 159, 163, 173
Negotiations
agreements, 166
Nigeria, 74, 76, 79, 81, 87, 88, 90, 137, 139, 141, 147
Nigerian Cities, 52–67

O
Optimum size, 90
Organization of the Petroleum Exporting Countries (OPEC), 139, 141, 143

P
Paris Agreement, 122–124, 129–130
Paris Rulebook, 158, 163
Petro-cities, 137–139, 141–146, 149–152, 160
Petro-dependency, 145–147
Petroleum industry, 139, 146, 149
Photovoltaic (PV) system, 74, 77, 80, 82–83, 85, 87–90, 97–100, 102, 105–107, 109, 110

Planning, 16–33
Policy
  brokers, 165, 168, 170–175
  formulation, 160, 163–165, 171–175
  instruments, 171, 176
  making agenda, 160
  policy-oriented learning, 166, 176
  subsystem, 160, 164–168, 171, 174
Political economy, 136–152
Pollutant emissions, 87, 96, 109, 110
Port Harcourt, 137, 141
Power converter, 77, 81
Public-private synergies, 145

R
Real estate, 28, 31
Reconfigurations, 136–152
Regulations, 17, 25, 26, 28, 29, 31, 32
Renewable Energy, 123, 126–131
Renewable fraction, 110
Renewable sources, 74, 88, 96, 97, 107
Rent-seeking, 146, 151
Resilience, 170, 171, 173, 174
Return on investment (ROI), 108
Rural-urban migration, 159, 170

S
Slums, 167, 176
Social mobility, 159
Socio-technical transition, 137, 138, 143, 144, 151
Solar photovoltaic (PV) system, 16, 17, 22, 24–32, 79–80
Stakeholders, 160, 163, 164, 170, 173, 175
State-of-Charge (SOC), 83, 86
Sub-Saharan Africa (SSA), 96
Supply chain, 52, 55, 57, 64, 67
Sustainability, 54, 67
Sustainable Development Goals (SDGs), 174, 176
Systematic review, 20

T
Techno-economic comparison, 85–88
Telecommunication sector, 90
Topology of cellular generation, 76
Top Runner, 61, 67
Transformations, 138, 146, 151
Transport, 52, 54, 61, 63, 64, 67
Tripoli, 137, 141

U
UNEP, 38–44
United Nations Framework Convention on Climate Change (UNFCCC), 120, 122, 123
Urban/urbanisation, 118, 119
  design, 17, 18, 27
  morphology, 17, 27–28

W
Waste
  management, 38–48, 53, 57, 58, 61, 64, 65, 67
  recycling, 42, 45
Waste to Energy, 46
Water electrolyser, 100, 101
Wind energy system, 77–79, 83
Wind turbine (WT), 78, 79, 81–89

The manufacturer's authorised representative in the EU is Springer Nature Customer Service Centre GmbH, Europaplatz 3, 69115 Heidelberg, Germany. If you have any concerns regarding our products, please contact ProductSafety@springernature.com

Printed and bound by CPI Group (UK) Ltd, Croydon, CR0 4YY
28/09/2026
02224072-0001